Plasma Processes for Semiconductor Fabrication

Plasma processing is a central technique in the fabrication of semiconductor devices. This self-contained book provides an up-to-date description of plasma etching and deposition in semiconductor fabrication. It presents the basic physics and chemistry of these processes and shows how they can be accurately modeled.

The author begins with an overview of plasma reactors and discusses the various models for understanding plasma processes. He then covers plasma chemistry, dealing with the effects of different chemicals on the features being etched. Having presented the relevant background material, he then describes in detail the modeling of complex plasma systems, with reference to experimental results. The book closes with a useful glossary of technical terms. No prior knowledge of plasma physics is assumed in the book. It contains many homework exercises and will serve as an ideal introduction to plasma processing and technology for graduate students of electrical engineering and materials science. It will also be a useful reference for practicing engineers in the semiconductor industry.

W. N. G. Hitchon received his B.A., M.Sc. and D.Phil. degrees from the University of Oxford and is Professor and Graduate Chair in the Department of Electrical and Computer Engineering and a member of the Center for Plasma-Aided Manufacturing at the University of Wisconsin, Madison. He is the author of over seventy journal articles and two previous books, *Flux Coordinates and Magnetic Field Structure* and *Semiconductor Devices: A Simulation Approach*.

T0181934

**Cambridge Studies in Semiconductor Physics
and Microelectronic Engineering**

TITLES IN THIS SERIES

PLASMA PROCESSES FOR SEMICONDUCTOR FABRICATION

W. N. G. HITCHON

CAMBRIDGE
UNIVERSITY PRESS

CAMBRIDGE UNIVERSITY PRESS
Cambridge, New York, Melbourne, Madrid, Cape Town, Singapore, São Paulo

Cambridge University Press
The Edinburgh Building, Cambridge CB2 2RU, UK

Published in the United States of America by Cambridge University Press, New York

www.cambridge.org
Information on this title: www.cambridge.org/9780521591751

First published 1999
This digitally printed first paperback version 2005

A catalogue record for this publication is available from the British Library

Library of Congress Cataloguing in Publication data

Hitchon, W. Nicholas G.
Plasma processes for semiconductor fabrication / W. N. G. Hitchon.
p. cm.
ISBN 0-521-59175-9
1. Semiconductors – Etching. 2. Plasma etching. I. Title
TK7871.85.H54 1999
621.3815′2 – dc21

98-11717
CIP

ISBN-13 978-0-521-59175-1 hardback
ISBN-10 0-521-59175-9 hardback

ISBN-13 978-0-521-01800-5 paperback
ISBN-10 0-521-01800-5 paperback

To my wife Jackie and son Adam.

To my wife Jackie and son Adam.

Contents

Ad majorem Dei gloriam

1

Introduction

This book is intended to introduce plasma processing and technology, so that the reader can readily understand the issues involved in processing and can immediately access the state-of-the-art literature. The reader is not assumed to have any knowledge of plasmas, but only some undergraduate background in basic electromagnetism. In fact, wherever possible the treatment even avoids quoting results from one part of the book for use in another part. In many cases it develops new ideas more than once, each time they are needed, to make the discussion easier to follow.

Plasma reactors and the physical and chemical processes that take place in them are discussed in considerable detail, the main emphasis being on capacitive, inductive, and electron cyclotron resonance (ECR) reactors. However, this is an area in which there are few simple (or perhaps even right) answers. It is not usually possible to give a complete description of a plasma reactor, so we are largely concerned, here, with showing how one should go about thinking about what is happening in the reactor.

The chapters of the book are organized in the order in which we need to consider material to understand the reactor. This can sometimes mean that we do not treat a single topic in one chapter separately from other topics. Instead, subject matter will often be introduced where it is needed as we build up the picture of how systems work. The discussion will move from an overview of processing systems based on simple arguments and considerations to more complex analyses as we progress.

Chapter 2 introduces the processes taking place in a plasma reactor, including the materials processing done by the plasma. This chapter describes the different levels at which the plasma may be understood; it touches on electrostatics, transport of particles (including random walks and "continuity" or "conservation" equations), and the role of feedback in the plasma.

The next two chapters are concerned (in Chapter 3) with the simplest plasma/ reactor models – circuit models – and (in Chapter 4) more detailed (mostly analytic) plasma models. Inductive, capacitive, and ECR reactors are all introduced and discussed in these chapters.

Chapter 5 begins to handle the specific issues that make processing plasmas particularly difficult to describe, with the topic of plasma chemistry. The chemistry, like some other aspects of these systems, is not well understood – much of the needed data is not available. The strategies used to deal with this uncertainty, and the "recipes" used in choosing plasma chemistries, are outlined.

The next two chapters extend the discussion of chemistry, relating the chemistry and long mean free path transport (since the pressure in many of these reactors is low), in Chapter 6, and the effects of the chemicals and their transport on the features being etched, in Chapter 7.

Chapter 8 returns to the hierarchy of means to describe the plasma, going into more detail than previous chapters. The rest of this introduction, and the final chapter, Chapter 9, are concerned with overall strategies for describing such a complex system and the results that can be expected from our attempts to do so.

Finally, a glossary of technical terms is provided, since much of the difficulty of learning a new field is caused by the unfamiliar expressions used to describe what are sometimes familiar ideas.

The remainder of the introduction is devoted to describing the approach that one might employ to study a physical system that is genuinely difficult to describe completely. It is an assumption that we make, in much of this book, that these plasmas are not capable of being described by means of simple analytic expressions. This may turn out to be untrue. Our approach is suitable for dealing with such a system – but it is also appropriate for finding an underlying simple pattern of behavior, within the complex behavior. In other words, whether the complexity of the system is real or apparent, the overall strategy we are advocating should be useful.

Strategies for a Complex System

This book is concerned with the process by means of which we can build up a useful understanding of a complex physical system. By "complex physical system" we mean a system that is sufficiently difficult to describe that it is not possible in practice to describe the system both completely and accurately. In such a case, there are few right answers to problems. The description of a complex system requires that the system be examined from many directions. Perhaps by starting with simple estimates, to show which effects are important, we can build enough understanding to construct more detailed calculations.

> With many calculations, one can win; with few, one cannot. How much less chance of victory has one who makes none at all! By this means I examine the situation and the outcome will be clearly apparent. [1]

Examples related to plasmas, used in fabrication of semiconductors, will illustrate a process where understanding of the plasma is developed, using whichever approaches are appropriate in a given instance.

> These are the strategist's keys to victory. It is not possible to discuss them beforehand.
> When confronted by the enemy respond to changing circumstances and devise expedients. How can these be discussed beforehand?

More important than the details of the way the estimates are done is the recognition that, by developing physical insight into the particular situation, a set of useful estimates and calculations can be arrived at, which will vary from one example to the next.

> Of the five elements, none is always predominant; of the four seasons, none lasts forever.

The purpose of these calculations is, ultimately, to be able to design a near-optimal system or process.

The overall strategy, used in designing the optimum process, must be developed with the widest possible view of the purpose of the design.

> Know the enemy and know yourself; in a hundred battles you will never be in peril.
>
> He whose generals are able and not interfered with by the sovereign will be victorious. The general should act expediently in accordance with what is advantageous and so control the balance.

Because the system we are focusing on is too complex to allow a complete description, there will be few neat, simple conclusions in what follows. We will not give the impression that we can divide the problem into separate topics that can be handled completely and separately, each in its own chapter. Instead, the chapters represent stages in the process of going from very simple estimates to complex calculations.

Before we can set up many different types of estimates of the behavior of our physical system, we need to have some idea of how to go about doing approximate calculations using the equations of applied physics. This is the topic of the next two sections. The last section of this chapter addresses the way in which such models are tested against experiments, in order to develop better models.

Problem Solving in Applied Physics

It is important to realize that problem solving is a skill in itself. Most people working in applied physics and engineering have more than enough knowledge to understand the material in this book. Using this knowledge to describe experiments needs something more than this.

The hardest part of solving many problems is the identification of the problem. Experiments raise many questions, and attempting to answer them is one of the main ways in which progress in science is made. Typically, the explanation of experimental observations in plasma processing of materials requires knowledge of the relevant physics, chemistry, and so on, and insight into which of the potentially relevant effects are actually important. For now, we will assume that these important effects have been identified. The task is then to describe them and their interaction.

The description of physical phenomena is usually done in one of two ways: by analogy (for instance, when we think of ions being like raindrops falling out of the plasma – see Section 2.1) or by means of mathematical equations. The challenge with analogies is to know when they break down. Equations could probably be said to embody a sort of analogy as well – and again, we need to know when they break down. The ability to formulate a description of either kind is very useful, and coming up with a good analogy can be most helpful in setting up an appropriate equation. The word "equation" means "setting equal," so when we write an equation we are describing two or more quantities, which are combined in the equation so that there is some kind of balance between them. Often, our task is to figure out which effects are "in balance," so we can write the equation that describes their balance. Once we have set up the equation(s), we will almost always be forced to solve them approximately. Exact solutions are rarely possible in practice.

One of the most frequently used equations in what follows is the continuity (or conservation) equation for the particle number density n:

$$\frac{\partial n}{\partial t} + \nabla \cdot \mathbf{\Gamma} = S. \tag{1.1}$$

This equation provides an excellent example of some issues which arise in the approximate solution of equations in applied physics. In steady state, $\nabla \cdot \mathbf{\Gamma} = S$ implies that S, which is the rate of production of particles per unit volume per second, is equal to $\nabla \cdot \mathbf{\Gamma}$, which is defined to be the rate they are lost from the surface of a unit volume per second. ($\mathbf{\Gamma}$ is the flux per unit area per second.) In the time-dependent version, given above, the rate of production S is balanced by the sum of two effects: the loss from the surface and the rate of increase in the density in the (unit) volume, $\frac{\partial n}{\partial t}$.

It is necessary to know when to use this equation, and what to use it to calculate. The short answers are: The equation is quite general (unless you can think of a term in the particle balance that has been left out) but it is only really useful if we know (expressions or values for) S and $\mathbf{\Gamma}$. It is normally used to find a particle number density, n.

These answers raise more questions: Why do we find n from it, not, say, S, and when do we know (expressions or profiles for) $\mathbf{\Gamma}$ and S?

The first part of the answer is that S is a "source" in the equation, which describes the production of density n. It is usual in such a case to assume we shall find n, from the equation and from knowledge of S. Other examples of equations with a source are Maxwell's equations, two of which read:

$$\nabla \cdot \mathbf{D} = \rho, \tag{1.2}$$

that is, the free charge density ρ "produces" the field $\mathbf{D}$, and

$$\nabla \times \mathbf{H} = \mathbf{J} + \frac{\partial \mathbf{D}}{\partial t}, \tag{1.3}$$

that is, the current density $\mathbf{J}$ "produces" the "rotation" of the field $\mathbf{H}$. In some practical examples it seems reasonable that we specify the source first. In an electromagnet, we "impose" a current density $\mathbf{J}$, and we calculate the magnetic field $\mathbf{H}$ it produces. If we create particles by some mechanism, in doing so we impose a rate of production S, and from this the density n must be deduced.

Reality is more involved than this, because the sources we impose are affected by the response of the system. Although this complicates the calculation, it does not change the fundamental relationship between the source and the response. (See Ref. [2].)

The conservation equation for the density is expressed in terms of the flux $\mathbf{\Gamma}$ and the source rate S. The quantities $\mathbf{\Gamma}$ and S can be found from the distribution function f, or from an infinite set of equations in terms of other "moments" of the distribution [3]. Moments of the distribution are integrals over velocity of f multiplied by integer powers of the velocity $\mathbf{v}$.

When the mean free path λ is small compared to other dimensions, we may be able to use approximate expressions for $\mathbf{\Gamma}$ and S, but these tend to be problematic, even then. If the mean free path is not small, we need to do a more accurate calculation based on a kinetic equation. The kinetic equation is a conservation equation, in many ways very like the equation for conservation of density. It uses a density per unit of "phase space" instead of a density in physical space. Phase space is a set of independent variables that combines physical space and velocity variables. Just like the more usual conservation equation, it has a source term, which is the rate of production of particles per second per unit volume of phase space. The density in phase space is the quantity we usually attempt to calculate when we use the kinetic equation, which is also what we should expect by analogy with the standard particle conservation equation.

Sensitivity to the Order of Calculation

The fact that the calculations we perform are approximate has an important consequence for the order in which we do calculations. If we know one quantity A, which appears in an equation, approximately, we may be able to find another quantity B in the equation, from our knowledge of A, with reasonably good accuracy. However, if we know B approximately we may not be able to find A from it at all accurately. The diffusion equation (1.1) illustrates this point well. In steady state, and if $\Gamma = -D\nabla n$, the equation is $-D\nabla^2 n = S$.

The particle number density n does not depend very sensitively on the ionization rate S. This means that if we know S approximately, we can find n from it with some confidence. However, an approximate knowledge of n would not be sufficient to allow us to find S with any confidence. One way to see this is to examine the analytic solutions of the diffusion equation given elsewhere in the book. The density is seen to tend to "fill-in" the center of a region, even when the ionization peaks far away. The density seems somewhat insensitive to the shape of S, although (for instance) doubling S in a linear problem will clearly also double n. Another way to look at this is to suppose that S had been measured and had errors of (say) 10%. This means that $\nabla^2 n$, the curvature of n has errors of 10%, but (in a plasma where $S > 0$) it is probably still negative everywhere. The density n itself when calculated may not be badly in error, because of a 10% error in its curvature. However, a 10% error in n, which could change from being $+10\%$ to -10% from one point in space to its neighbor, could make both ∇n and $\nabla^2 n$ completely wrong – and could for instance easily make them both have the wrong sign. If we find S from n, by setting $S = -D\nabla^2 n$, then S will similarly be quite wrong.

As a consequence of this, we should set up calculations where we iterate to find the profiles of n and S, so that we use the diffusion equation to find n from S. We must use another, direct calculation to find S – that is, we must use a method directly related to the way in which the ionization takes place. This might involve calculating the heating rate in the plasma and then determining how much of the power deposited is used for ionization, for instance.

In summary, the (partial differential) equations we use often have a "source" and a "response." We usually should specify the source, and calculate the response from it. This is physically reasonable; it is also the case that the response is less sensitive to the exact details of the source, than the source would be to the details of the response, if we tried to use the equation the "wrong way round." This is important, given that our knowledge of any variable is usually approximate at best. Mathematically, the source usually appears directly in the equation, whereas the response is usually differentiated one or more times with respect to space and/or time. The differentiation of the response makes the equation very sensitive to the details of how the response varies in space or time.

Design of Calculations

In a text on statistical design of experiments [4], some guidelines are given that, with very slight modifications, are extremely useful in designing the calculations we need to do.

We expect to perform a series of calculations, where we speculate about the nature of the process we are studying in order to describe the process. We compare the results of our calculations to experiments, to see whether we successfully included the important aspects of the process. The results of the comparison should lead to improvements in our speculations. It may also suggest new experiments. The cycle of comparison and improvement is expected to be iterated multiple times.

Understanding the nature of the problem and the variables involved requires careful consideration and the use of all the available information about the process. This aspect of the procedure resembles Sun Tzu's approach to a military campaign, where the importance of: 1) recognizing what is really important, 2) seeking out information, and 3) making estimates of various kinds, are stressed.

Montgomery [4] summarizes the stages of designing a statistical experiment, in a fashion we shall rewrite to make it apply to our situation, as:

1. Use your knowledge of the problem to determine which effects are likely to be important.
2. Keep your analysis as simple as possible. Including the important effects is what is important, rather than using an enormously complex model.
3. Recognize what is actually significant in practice. Do not get bogged down in minor details.
4. Iterate. We usually learn what is most important as we go along. Often one should not invest more than 25% of one's effort in the initial calculations, which are frequently just learning experiences.

These points should be illustrated with examples. Taking them in order, we have:

1. Many people put a great deal of effort into the wrong aspect of a model, and they fail to recognize that other aspects of the problem are more critical. Most plasmas are controlled to a great extent by surface processes. There is little point having an exact plasma model, if we fail to properly consider how the surface processes control the plasma.
2. Complex plasma models are often set up, with a "module" to describe each aspect, but only some of the modules are critical. It is more important to get the critical aspects right, than to have a module for every conceivable process.
3. One of the common mistakes, made most frequently, but by no means only by inexperienced researchers, is to find a very small error in an existing calculation

and put a great deal of effort into correcting it. Suppose we can do calculations, or make measurements, to an accuracy of 5%. Suppose further that we find errors in our calculations, which can be corrected with a certain amount of effort. Suppose there are three errors, as follows:

	Size of Error	Time to Correct Error
Error 1	50%	1 month
Error 2	5%	1 hour
Error 3	5%	6 months

A surprising number of people will simply decide to fix all three errors, and do them in no particular order, when in fact error number three should certainly be tackled last, if at all.

4. Frequently, we do not know which effects to include. In one situation, it took many years of estimating the magnitude of different heating processes to discover that Coulomb collisions were effective in heating "cool" electrons in a dc discharge. For this reason, we need to iterate – we could not have set a correct calculation up at the outset. It took many iterations to rule out some processes and hence to deduce which process was missing.

Exercise

1. What are the major considerations in choosing a plasma reactor for etching silicon-based wafers? Try to be realistic, for a) a new industrial plant and b) a small-scale industrial or university project. What typically determines the diagnostics and modeling that are employed? Analyze the cost of adding various diagnostics, and employing models, to optimize the process. Discuss whether it is economically justified to increase the use of diagnostics and modeling.

2

Plasma Processes for Semiconductors

The plasmas used in processing semiconductors are usually partially ionized gases, the neutral gas pressure being from about 1 mTorr to about 100 mTorr and the plasma (that is, the charged particle) number density n being in the range 10^{10} cm$^{-3} \lesssim n \lesssim 10^{12}$ cm^{-3}. To describe these plasmas precisely and in detail is difficult for a number of reasons. Plasma physics textbooks are largely devoted to detailed analysis of the special cases that can be described by analytic theories. Instead, it is often more useful to try to obtain approximate descriptions of the real plasma using simple physical reasoning.

In trying to understand a complex physical system it is usually useful to try several different approaches, such as studying different physical pictures, and to attempt to reconcile them with available data. In this book we will attempt to describe each situation we consider starting from "first principles" in each case. A major theme throughout will be the development of approximate quantitative models to explain experiments, based on simple reasoning. Statistical design and analysis of experiments, for example [5], provides a valuable tool for building up physical understanding.

It is tempting to say that theoretical models of processing plasmas are like houses made of cards. After a few layers (of assumptions) the whole edifice is in danger of collapsing. A slightly better metaphor might be a house of straw, where each extra layer (assumption) we add bends the structure so that after a very few layers it is hopelessly distorted.

For this reason it is important to avoid taking a simple analytic model and pursuing it as far as we possibly can. Instead, we need to be aware of the analytic models that might be relevant and that might explain experimental data, and we need to constantly check them for realism when we try to use them. Space will not permit this testing process to be illustrated fully in what follows, but it is important to keep in mind the necessity for vigilance in this regard. Typical assumptions we will try to avoid, but which are commonplace in some relevant literature, are that inelastic rates of reaction can be found assuming a Maxwellian electron distribution or that the transport of particles (neutral or charged) at long mean free path can be described by "fluid" equations.

2.1 Plasmas

The plasmas we are interested in are a particular subset of all the possible types of plasma: Processing plasmas are partially ionized gases. In general, the definition of a plasma could be given as: A plasma is a collection of charged particles where the long-range electromagnetic fields set up collectively by the charged particles have an important effect on the particles' behavior. In our case, the fields the plasma sets up will be mostly electric fields. Further, we are concerned with plasmas confined in "tanks," and the electric fields are usually strongest where the plasma is in contact with the surface of the tank. This electric field is created because electrons in the plasma tend to move much faster than ions. The fast-moving electrons hit the wall before the ions do and some stick to it, so the wall charges up negatively and this negative charge pushes other electrons away at the same time as attracting positive ions (see Fig. 2.1). In steady state the wall no longer charges up, and thus electrons and positive ions have to arrive at the same rate. The field near the wall holds the electrons away from the wall, allowing only the most energetic electrons to get there. The field also accelerates the positive ions toward the wall, and in this way the rates of arrival of electrons and positive ions are made equal.

The electrons on the wall push the other electrons back from the wall, exposing the positive charge in a region close to the wall. This positively charged region is called a "sheath." The positive charges in the sheath are the starting points for electric field lines. These field lines end on electrons on the wall. The electric fields in the positively charged sheath region are very strong. The interior of the plasma is usually nearly neutral and the electric fields in the plasma interior are much weaker. The strong electric field in the sheath accelerates the ions in a direction straight toward the surface. The high-energy ions moving nearly normal to the surface are a vital aspect of the plasma processing of materials.

The positive ions are produced somewhere inside the plasma (as are most of the electrons) and fall outward (see Fig 2.2). Like raindrops steadily falling from a cloud, the number of ions or raindrops falling through an area per second (in steady state) is the total number created each second above that area. The speed of the ions, or drops, depends on how far they have fallen, and on whether they hit anything while they were falling. The density of ions, or drops, will be greater if the number passing per second is high and if they move slowly.

In this analogy the cloud corresponds to the source of ionization. In the plasma, gravity is not usually important. It is the electric field that tends to pull ions out. The field usually points outward from the "center" of the plasma where the field is zero. The ions passing through a plane are those produced between the center where the field passes through zero and that plane. The ion density and the number crossing an area per second depend on where the ionization is, on where the electric field goes to zero, and on how the field varies in space.

Another metaphor for the ion motion, in contrast with the electron motion, is that the ions are as slow as molasses running downhill. The plasma potential appears to the positive ions like a hill, with the peak of the hill being somewhere in the middle of the plasma volume. The molasses (ions) is being poured onto the hill around the peak and moves very slowly outward.

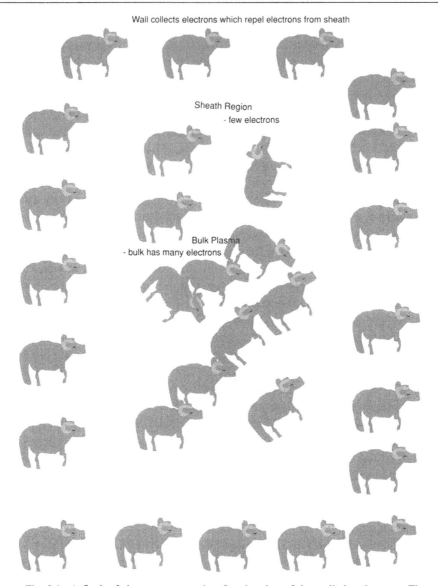

Fig. 2.1. A flock of sheep as a metaphor for charging of the walls by electrons. The fast-moving electrons hit the wall and stick to it, providing a negative wall-charge. The negative charge initially builds up faster than positive ions can get to the wall to neutralize it. In steady state the negative wall-charge repels other electrons from the region close to the wall (called the sheath) and attracts ions, so the fluxes of electrons and ions can be equal.

The electrons behave somewhat like sheep in a croft. See Figs. 2.1 and 2.3. (A croft is a small field; it might be confusing to talk about the sheep being in a field.) The croft slopes a little upwards from the point of view of the sheep (electrons) as they move out from the middle. There is a very high fence (the sheath) around the croft. Very few sheep can leap over the fence. Some sheep are so weak that they are trapped in the center of the croft by the slope, but many of them are energetic enough

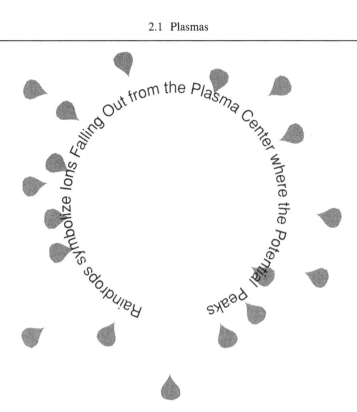

Fig. 2.2. Rain as a metaphor for the ion flux. Ions are produced in the interior of the plasma and fall outward, under the influence of the electric field, which tends to pull ions toward the nearest wall.

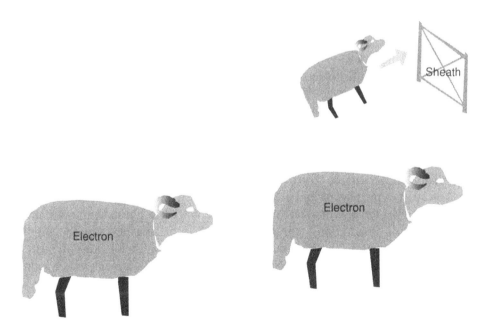

Fig. 2.3. Sheep as a metaphor for electron behavior. The electrons can escape to the wall by getting over the potential barrier provided by the sheath.

that they run around the croft and try to escape when they reach the fence. Only very occasionally can one of them get over the fence. If there is a low point on the fence, nearly all the sheep that do escape will escape at the low point.

If part of the boundary is a good conductor, the top of the fence is at a uniform height along that section of the boundary. If the boundary has portions that are insulating, the height of the top of the fence can vary along the insulating portion. This may allow the sheep to find a low point at which to escape. However, if they do escape easily to some point on the insulator, the rate at that point is likely to be faster than ions are escaping to any point. The fence height will rise there. In fact, the whole fence is actually a pile of sheep that ran to the edge of the croft and stopped there. (The sheath is set up by electrons on the wall, which repel other electrons.)

The application of an alternating voltage to part of the boundary raises and lowers the fence. Sheep (electrons) surge toward a piece of boundary where the height has been lowered, even if they cannot get over the top. If the barrier is raised, the sheep surge away.

2.2 Plasma Interactions with Materials

Plasmas are useful in semiconductor processing primarily because the plasma electrons are hot and can drive chemical and other processes that would not be encountered in equilibrium, and because the positive ions arrive at the surface with a high velocity directed nearly normal to the surface. The ions can, among other things, sputter (that is, knock) material off the surface, damage the surface, provide heat to the surface, or implant in the surface. Because the sheath electric fields cause ions to arrive at near normal incidence they are much more likely to hit the bottom of a trench than a vertical sidewall of a trench, and this is crucial for the faithful transfer of a pattern during etching. Plasma ions striking a semiconductor through a hole in a mask will tend to dig the hole straight down, whereas chemical etching will dig sideways into the semiconductor as well as down. The electrons ionize neutrals, break molecules apart to form radicals, create molecules in excited states, and heat the surface. The radicals are frequently responsible for etching the surface. Radicals may also polymerize on the surface and form a layer that protects the surface.

In deposition processes such as Plasma Enhanced Chemical Vapor Deposition (PECVD) the electrons create species in the gas phase that participate in the film deposition on the surface. During plasma polymerization the original molecule is broken apart in the plasma and polymerizes to form a film, although it is often not known whether the polymerization takes place in the gas phase, on the surface, or in both places. In addition, "polymers" formed in plasmas are usually complex, cross-linked structures that lack the repetitive chain structure of true polymers.

Etching discharges may achieve the etching by ion bombardment, which sputters the surface, or by chemical etching. Sputtering (Fig. 2.4) is the process where ions knock atoms off a surface and the atoms deposit on another surface. Metal is sputtered to make electrodes and interconnects. A very common approach is to use Ar plasma to supply ions with energies of up to several keV. Sputtering is often done in "magnetrons" [6, 7], in which a magnetic field is used to help keep plasma

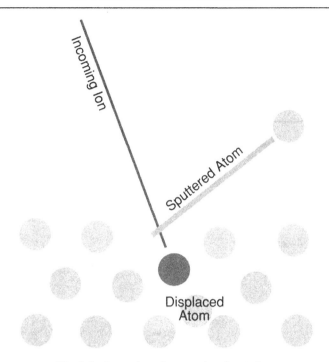

Fig. 2.4. Sputtering of a target by plasma Ions.

electrons inside the discharge. If the target being sputtered is not a single element, there can be an initial transient in which a light element comes off the surface "early." Eventually the sputtered material has to have the same composition as the target, however.

If ε_b is the surface energy binding the atoms in the target (which is a few eV) the number of atoms coming off the surface with energy ε is proportional to $\varepsilon_b \varepsilon / (\varepsilon_b + \varepsilon)^3$.

The deposited film's morphology has been described empirically by the "Thornton" diagram [8], which is an extension of the "Movchin–Dimchishin" diagram [9]. The film is affected most by (i) the energy of the bombarding ions and (ii) the substrate temperature (measured relative to the melting temperature), both of which provide activation energy for the growing film to rearrange itself. If sputtered atoms (adatoms) stay where they first land on the surface, columns of material grow, with gaps between them where the columns cast "shadows." As the substrate temperature rises, or the ion bombardment energy increases, the adatoms are given activation energy that allows them to diffuse across the surface and fill in the voids between the columns. At high temperatures the material is able to "relax" into grains that are the material's equilibrium configuration. (See [10].) Modern computational studies of these aspects of surface behavior tend to employ molecular dynamics [11] methods, using accurate interatomic potentials [12, 13].

Ions from a plasma are sometimes implanted into a surface [14–16], for instance to "dope" the material. This process has been modeled and discussed in Ref. [17]. Data on ion transport and scattering in solids are given in Refs. [18]–[27].

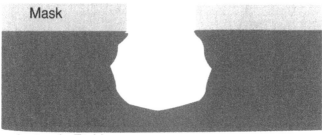

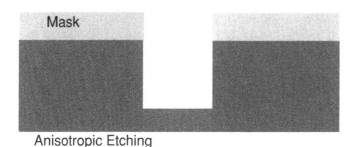

Fig. 2.5. Anisotropic versus isotropic etching.

Chemical etching by itself is often not desirable because there is no preferred direction – purely chemical etching is "isotropic" (Fig. 2.5). Etching by ions takes place by sputtering and is anisotropic since it has a preferred direction – and the fact that ion etching is anisotropic is desirable. However, sputtering by ions will etch most materials, which is not ideal. It is desirable for the etching to be selective and etch some materials rather than others. Chemical etching, by contrast, is capable of having high selectivity. To achieve both selectivity and anisotropy simultaneously, etching processes are usually designed to employ both chemical etching and the physical etching that is done by the ions.

A commonly used example of the combination of chemical and physical etching is etching in the presence of a gas that will polymerize on the surface. The polymer formed from the gas protects the surface from chemical attack. Plasma ions sputter off the polymer and other material and expose the surface to chemical attack. Anisotropy is achieved because ions only clean the polymer off the bottom of a trench and not off the side walls; consequently, the bottom is etched but not the sides. Selectivity might depend on the chemical etch of the bare surface only being effective for (say) Si or SiO_2. The selectivity might also be achieved because the polymer may only form on one or the other, of the two materials. In etching using CF_4 gas a polymer is formed on the surface of Si but not SiO_2. The SiO_2 gives up oxygen to form CO_2 from the radicals, which contain carbon (and which would otherwise form a film). This prevents the formation of the polymer on SiO_2 and so SiO_2 will be etched while Si is protected from etching by the polymer [28].

To damage a solid material by particle bombardment about 20 eV of energy per particle is needed to knock atoms of most materials off their sites in the lattice. Ions striking surfaces after falling through a sheath usually have at least this much energy, unless energy is lost in collisions that occur while the ion is crossing the sheath. Ions are consequently capable of damaging the substrate.

Rather little is known about the processes that take place on the walls in a plasma environment. As an example, in experiments in CF_4 discharges there is evidence that although CF_4 molecules are rapidly broken up by electrons in the gas phase there is a great deal of CF_4 gas present. It is believed that CF_4 must be reforming from its constituent parts on the walls. However, little is known about the process at the wall or about what mixture of products we should expect to come off the wall in different circumstances. This situation is typical of most plasmas used in materials processing.

2.3 The Hierarchy of Plasma Models

In this section we examine the physical models needed to build up an overall picture of the plasma. The example we use in this section is a type of device that uses an "Inductively Coupled Plasma" or ICP. The overall approach to understanding other types of plasma reactor is very similar to that which is outlined here. Our preliminary discussion of the ICP is intended to show the role of the many different models we use to describe processing plasmas.

The ICP is rapidly becoming one of the most important types of plasma reactor used in semiconductor processing. A common configuration consists of a cylindrical vacuum tank with the axis of the cylinder being vertical, in the z direction. There is a circular quartz window inside the cylinder, at $z = z_o$ at the top of the plasma volume. Just above the window is the antenna used to create the plasma. The antenna usually consists of a spiral coil or some similar arrangement.

The antenna is the primary coil of a transformer, which is supplied with radio-frequency power. The plasma forms the secondary. This implies that the first model we should employ to describe the ICP is a circuit model. It should be clear that the information we can get from the circuit model is limited in its scope, since it will not address conditions inside the plasma, for instance. It is also limited by the accuracy with which we can represent the ICP in terms of simple circuit elements. From the point of view of describing the coupling of the external power supply to the plasma, however, the description of the plasma "circuit element" is a crucial step.

When we try to describe this "transformer," we have to consider the currents that will flow in the plasma. The current is dependent on the conductivity and this in turn depends on the density and on the collisions of the electrons (since the electrons carry the vast majority of the radio-frequency currents in the plasma). The response of the plasma to the magnetic field set up by the primary is critical, not only because it determines the performance of the transformer, but also because the amount and location of the power deposited in the plasma is determined by it. The second type of model we are led to consider is thus one that deals with the electromagnetic fields set up by the primary and the plasma.

As we follow this line of thought further we are forced to look into the way we can find the conductivity. The conductivity depends on (for instance) whether the electrons in the plasma are suffering frequent collisions or not. To assess how the electrons are behaving we estimate the time scales of the most important processes affecting the electrons. The third class of model, which we are examining now, involves the particle kinetics taking place in the plasma and leads to considering the way particles move around inside the plasma.

Transport processes make up a topic we will go into in some detail since some of the concepts need explaining in detail and since understanding these processes is vital if we hope to understand the plasma. The particles in the plasma exhibit an interesting range of different types of behaviors as they move through the chamber. The positive ions in the plasma usually fall outward from the plasma toward the walls, pulled by the electric fields. They are "born" nearly stationary when a neutral atom or molecule is ionized and they pick up more speed the further they go, except when collisions slow them down or deflect them. Where they are born and the electric field through which they fall determine where they hit the wall.

Electrons are confined inside the plasma by the electric field, except for the most energetic of them, which can get to the walls. They usually have many elastic collisions in the time it takes them to cross the plasma volume; and they also tend to have many elastic collisions in the typical time between inelastic collisions. Elastic collisions are those collisions where no kinetic energy is used in changing the internal state of the particles – that is, exciting or ionizing them as occurs in an inelastic collision. Kinetic energy is exchanged between particles in elastic collisions, but when a particle bounces off something (stationary) which is much heavier than itself it only gives up a little energy. This all adds up to a picture where electrons typically move around having multiple elastic collisions, changing direction in each elastic collision but not changing their kinetic energy much in each elastic collision. Less frequently they bounce off the strong electric field near the chamber wall. If they are relatively energetic they may escape to the wall, or they may have an inelastic collision and lose a considerable amount of energy. The elastic collisions cause the electrons to undergo a "random walk."

The random walk means that if the electrons travel from point A to point B but have several elastic collisions in between they will take a path that is considerably longer than the minimum distance from A to B, with a consequent increase in the likelihood of an inelastic collision if they are energetic enough to excite or ionize a neutral. The electrons are usually responsible for the ionization that sustains the plasma, and in steady state the rate of ionization must exactly balance the loss rate. Every electron must, on average, produce one other electron before it escapes to the wall. To reach the wall the electron must get over the electrostatic potential barrier that is pushing it away from the wall. However, only the most energetic electrons can escape to the wall or can ionize a neutral to produce other electrons.

The electrons and ions in the plasma also interact with the neutral gas to produce neutral species, some of which would be highly unlikely to form in the absence of the plasma. Chemistry in the gas phase is closely linked to the transport processes taking place. The electrons, ions, and neutrals bombard the surface and perform the processing of the surface material, which we are attempting to understand, to control, and to optimize. The interaction of the gas with the surface is the next, and most important, topic to be addressed.

The species from the plasma attack the surface physically by hitting it and transferring kinetic energy, heating it, and possibly knocking atoms in the solid out of their site in the solid structure. The plasma species also react chemically with the surface. Charged particles charge up the surface. The physical and chemical attacks on the surface will often combine to remove material that neither could remove alone. For instance, ion bombardment can "sputter off" a layer of polymer from the surface, exposing the material beneath to chemical attack. When atoms deep in the solid are knocked off their site the solid is damaged. Damage can also be done when the surface charges up too much, causing a large electric field, so that if there is an insulating layer next to the surface it breaks down. Ions are attracted to the surface by the electrons that are already on the surface; thus they tend to hit the surface while moving in a direction that is nearly straight toward the surface. This means that if a trench is being cut in the surface, the ions are more likely to hit the bottom than the sides, and this helps to etch the bottom faster than the sides are etched.

Some of these surface interactions can be described using straightforward physical arguments, but most of what happens on the surface is not well understood at present. Our fourth set of models, which attempt to describe the surface, will include some elements based on first principles arguments, but part of our strategy must be to make even more use than usual of suitable experiments to guide our thinking.

In this section we have introduced four major categories of physical pictures of different aspects of the plasma reactor:

1. circuit models of the external circuit and the plasma;
2. electromagnetic descriptions of the coupling of power into the plasma;
3. transport and other gas phase processes in the plasma region itself; and
4. surface interactions.

Other subsystems will be investigated as they are needed. A brief introduction was given to indicate the scope of the physical models employed and to start the process of thinking in appropriate physical terms about each of them. This approach will soon provide the overview of the reactor that we need to make sense of the reactor's behavior. In Chapters 4 and 8 we shall discuss the role of analytic models and computational solution of the equations [29] used in describing the reactor.*
The intent of this approach is to establish a clear picture of the major issues. Armed with this, we should be in a stronger position to turn to the literature for information on details [30–33].

Other useful texts on plasmas include Refs. [6], [7], [34]–[44] (see particularly the article by P. C. Johnson in Ref. [34]) and the classic by von Engel [45]. References [46], [47], and [7] have more emphasis on processing. Books on fluid dynamics [48–51], on chaos [52], on mathematical methods [2, 53], and on computational methods [54, 55] are also likely to be useful.

*See also our web site at clerc.engr.wisc.edu.

The rest of this chapter will concentrate on explaining the overall issues involved in the physical description of the plasma, which correspond to items 2 and 3 above. The first topic will be electrostatics in a plasma; this is followed by an introduction to transport processes.

2.4 Plasma Electrostatics

One of the aspects of a plasma which is most difficult to describe is that the electric (and magnetic) fields in the plasma are set up self-consistently by the plasma, and the plasma is controlled by those fields. Even the simplest description of the plasma in the self-consistent electric field (SCEF), the fluid description, is remarkably complex.

First we should establish the conditions when the electric field in the plasma is strongly affected by the plasma. Poisson's equation can be written, in one dimension, as

$$\frac{dE}{dx} = \frac{\rho}{\varepsilon},$$
(2.1)

and so the change in the electric field in a distance Δx is of the order $\Delta E \sim \frac{\rho \Delta x}{\varepsilon}$. If we know the size of the vacuum electric field, E_{vac}, we could compare this ΔE to it. If $\frac{\Delta E}{E_{\text{vac}}} \ll 1$ the plasma has little effect. If we consider an extreme case, which is in practice a sheath region where $\rho = ne$, that is, there are ions but no electrons, then this is equivalent to a condition on n in terms of E_{vac}, $\frac{ne\Delta x}{\varepsilon E_{\text{vac}}} \simeq 10^{-8} \frac{n\Delta x}{E_{\text{vac}}} \ll 1$, where n is the number of ions per cubic meter. However, it is easier to estimate vacuum voltages than to find E_{vac} in most cases.

Poisson's equation can also be written as

$$\frac{d^2\Phi}{dx^2} = -\frac{\rho}{\varepsilon}.$$
(2.2)

Thus $\Delta\Phi \sim \frac{\rho(\Delta x)^2}{2\varepsilon} \simeq 10^{-8}n(\Delta x)^2$. If this is much less than the voltage in the absence of the plasma, then the plasma has little effect on the electric field. If the expected voltage is about 100 volts, and the chamber size is about 10 cm or 0.1 m, then if $\Delta\Phi$ is to be 100 volts,

$$100 \simeq 10^{-8}n(10^{-1})^2$$
(2.3)

and $n \simeq 10^{12}$ m^{-3} or 10^6 cm^{-3}. In contrast, in a "chamber" of size 10^{-6} m (in other words inside a feature being etched in a piece of silicon), then to get $\Delta\Phi \simeq 10$ volts we need

$$10 \simeq 10^{-8}n(10^{-6})^2,$$
(2.4)

so $n \simeq 10^{21}$ m^{-3} or $n \simeq 10^{15}$ cm^{-3}.

In summary, the plasma number density in a plasma reactor has to be in excess of 10^6 cm^{-3} for there to be a "real" plasma, which can affect the electrostatic potential Φ, in the main chamber. This is a very low density compared to densities used in plasma processing. In a trench, however, the number density is highly unlikely to be high enough to affect the electric field, since it would need to be 10^{15} cm^{-3} to do this.

The usual situation in a plasma reactor is that the density of ions and electrons is high enough to shield out a typical applied voltage in a very short distance, so that the plasma interior can be nearly free from strong electric fields, and the electrons can provide a charge density that nearly neutralizes the plasma in the interior. If the number density of ions at the plasma edge, where most of the shielding takes place, is about 10^{10} cm^{-3} or 10^{16} m^{-3}, then to shield out 20 volts we need a thickness Δx given by

$$20 \simeq 10^{-8} n (\Delta x)^2 = 10^8 (\Delta x)^2. \tag{2.5}$$

Thus $(\Delta x)^2 \simeq 10^{-7}$ m^2 and $\Delta x \simeq 3 \times 10^{-4}$ m.

This implies that the electric field is large in a very narrow sheath, at the edge. Electric field lines start on positive ions in the sheath. The field lines end on electrons on the wall. The rest of the plasma is nearly neutral, in the sense that if we know the ion density n_i, then the electron density n_e is close to it. (This does not mean that the charge density $\rho = e(n_i - n_e)$ is close enough to zero to set $\rho = 0$ in Poisson's equation.)

To estimate Φ in the plasma interior, we base our thinking on the fact that the electron fluid is nearly in equilibrium in the bulk of the plasma. As a result, the electron fluid in a plasma usually carries a lot less current than it could if all the electrons were going in the same direction. The "fluid" equation for the electron flux per unit area has two terms. To make the current small, these two terms must nearly cancel. The flux per unit area is discussed in Sections 2.6 and 2.7; it is

$$\mathbf{\Gamma} = -n_e \mu_e \mathbf{E} - D_e \nabla n_e \simeq 0 \tag{2.6}$$

(where μ_e is positive and the negative sign associated with μ_e is shown explicitly.) This has the equilibrium solution

$$n_e = n_o \exp \left[\frac{\mu_e \Phi}{D_e} \right], \tag{2.7}$$

which is the Boltzmann expression for the density. When n_e is set equal to the ion density n_i, the potential is found to be

$$\Phi = \frac{D_e}{\mu_e} \ln \left[\frac{n_i}{n_o} \right] = \frac{k_B T_e}{e} \ln \left[\frac{n_i}{n_o} \right], \tag{2.8}$$

since $D_e = \frac{k_B T_e}{e} \mu_e$.

If the electrons are truly at a temperature T_e then the potential can be found from this.

One interesting feature of this Boltzmann potential is that it tends to "heal" the ion density. It tries to flatten peaks in the ion density (including the main peak in the middle of the plasma) and it tries to fill in holes. The electric field is

$$E = -\frac{d\Phi}{dx}; \tag{2.9}$$

so

$$E = -\frac{k_B T_e}{e} \frac{1}{n_i} \frac{dn_i}{dx}. \tag{2.10}$$

If n_i decreases with x then $\frac{dn_i}{dx} < 0$ and positive ions are pushed to larger x by E, which is positive. If $\frac{dn_i}{dx} > 0$ ions are pushed to smaller x.

2.5 Transport of Particles and Energy

Transport theory is a topic critical to this work and thus we need to review it carefully. The concepts we shall discuss are probably familiar ones; we need a thorough understanding of the simple ideas to be able to set up appropriate descriptions of the plasma. One of the most important issues is the random walk of particles in the gas phase that occurs in the course of a series of collisions with other gas particles.

2.5.1 Gas Phase Collisions

Many of the particles we are concerned with undergo a "random walk" (see Ref. [56]) as they move around. The "steps" of the random walk take place between collisions with other particles (or with the walls, which for electrons usually means with the sheaths). Before we discuss the random walk we shall calculate the probability of a collision.

Typically we assume that the particles collide with other, "background," particles. The number per unit volume of these other, background particles is N, and we assume that the background particles present a "cross section" σ to the particle, which we shall call a "test" particle to distinguish it from the background particles. As the test particle approaches a background particle, it will collide with the background particle if the center of the test particle passes through the area σ around the center of the background particle. If we ask how far it can go before colliding, we can answer this by imagining the test particle moving in a straight line and hitting any particles whose centers fall inside an area σ centered on that line. (If the test particle's center has to be inside an area σ centered on the background particle's center, then the background particle's center also has to be inside an area σ around the center of the test particle.) In this way the test particle sweeps out a volume equal to σ times the distance traveled and hits any particles with their centers in that volume.

If the particle travels on average a mean free path λ before colliding, then it sweeps out a volume $\lambda\sigma$ on average before having a collision. If it has on average one collision when it sweeps out the volume $\lambda\sigma$, and it collides with all the particles with their centers in that volume, there must on average be one particle in the volume. Because the number density is N, the volume that contains one particle is $1/N$, and so

$$\lambda\sigma = 1/N; \qquad (2.11)$$

and

$$\lambda = 1/N\sigma. \qquad (2.12)$$

This is only the average distance traveled between collisions of a given type. The distances traveled by particles between collisions are distributed from zero to infinity (or the distance to the chamber wall).

There are many kinds of collisions and we need to be aware of the mean free path for all of the important ones (which includes the kinds that have "small" λ). The mean free path λ usually depends strongly on the kinetic energy of the particle, because σ depends on the kinetic energy.

The collision time τ_c, which is the average time between collisions, is just the time to go a distance λ,

$$\tau_c = \lambda/v, \tag{2.13}$$

where v is the particle speed. Usually τ_c and λ vary significantly with kinetic energy but occasionally one or the other will be roughly constant in some range of energies. The collision frequency $\nu = 1/\tau_c$.

2.5.2 A Simple Random Walk

To understand the nature of the random walk it is useful to consider a situation where the particles travel a distance of exactly λ and then collide, instead of only traveling λ on average. To simplify matters further, consider a one-dimensional random walk in x where all the particles we are interested in start at the same place, $x = 0$.

After one step, all our particles move out from $x = 0$ to $x = \lambda$ or $x = -\lambda$. In the second step, on average we expect half of the particles at $x = \lambda$ to take a step of $\delta x = +\lambda$ and go out to $x = 2\lambda$, but the other half to take a step of $\delta x = -\lambda$ and go back to $x = 0$. A similar splitting happens to the particles that had gone to $x = -\lambda$, so that after two steps the particles are expected on average to be arranged as follows:

A quarter at $x = 2\lambda$.
A half at $x = 0$.
A quarter at $x = -2\lambda$.

The fastest that any particle can move away from $x = 0$ is at a speed $v_{max} = \lambda/\tau_c$, which it does if its steps are always in the same direction. But the number of particles that have always gone straight out drops by on average a half at each step, so although no particles can go farther in N_s steps than $N_s\lambda$, very few particles go that far if N_s is large.

In a random walk, the mean square distance the particles have gone does increase linearly with time (or with the number of steps, which is equivalent), but again this is only on average. The mean square distance

$$\overline{x^2} = \left(\sum_{\text{all particles}} x^2 \right) \Big/ (\text{number of particles})$$

$$\simeq \lambda^2 N_s. \tag{2.14}$$

If we use the square root of this to measure the typical distance the particles have gone, then the root mean square distance x_{rms} is

$$x_{rms} \simeq \lambda\sqrt{N_s}. \tag{2.15}$$

What this means is that if we start a large number of particles at $x = 0$, then after N_s steps they will have spread out by an average distance $x_{rms} \simeq \lambda\sqrt{N_s}$. Some will have gone further (some will have gone a lot further) but most will probably have gone less far than x_{rms} and the highest density will probably be near $x = 0$.

2.5.3 Random Walks of Electrons in a Low-Pressure Plasma

To give a concrete example of why random walks are important, which illustrates several related concepts, we can try to understand the motion of the electrons in a low pressure (meaning low neutral-pressure) plasma. The electrons need a lot of kinetic energy to ionize a neutral – more than most of them have. Even if they do have that much kinetic energy the mean free path that an electron travels before ionizing a neutral, λ_e^{ion}, is often quite large. Suppose the chamber containing the plasma was a cylinder with a radius of $R \simeq 30$ cm and a height L comparable to this, $L \simeq 30$ cm. Suppose λ_e^{ion} was somewhat bigger, $\lambda_e^{\mathrm{ion}} \simeq 50$ cm but that the electron mean free path for elastic collisions λ_e^{el} was smaller, $\lambda_e^{\mathrm{el}} \simeq 5$ cm. If the electrons are heated at the top of the chamber in a range of radii from $r = 15$ to $r = 20$ cm (corresponding to an inductive heating coil near $r = 18$ cm, for instance), where will the ionization take place?

By the time the electrons have traveled $\lambda_e^{\mathrm{ion}} = 50$ cm, a lot of them will have ionized a neutral. In figuring out whether they have gone far enough to have a chance to ionize, it does not matter whether they went in a straight line for half a meter or went backwards and forwards for $N_s = 10$ steps of $\lambda_e^{\mathrm{el}} = 5$ cm. The electrons are unlikely to have gone in the same straight line, because elastic collisions make the electrons have a random walk with (average) step size λ_e^{el}. When the number of steps of that random walk, N_s, is large enough the electrons will have gone far enough for many of them to have ionizing collisions. This happens when the total distance traveled $N_s \lambda_e^{\mathrm{el}}$ is equal to the mean free path for ionization, or

$$N_s \lambda_e^{\mathrm{el}} = \lambda_e^{\mathrm{ion}}. \tag{2.16}$$

It is likely that most of the distance traveled was in a backwards and forwards motion (see Fig. 2.6) so the electrons will have gone less far than this away from where they started. The typical distance they will have moved outwards is

$$x_{\mathrm{rms}} = \lambda_e^{\mathrm{el}} \sqrt{N_s} \tag{2.17}$$

since λ_e^{el} is the step size of the random walk between elastic collisions. Now $N_s \simeq \lambda_e^{\mathrm{ion}}/\lambda_e^{\mathrm{el}}$ by the time an ionizing collision becomes likely. This means the mean distance they go before ionizing a neutral is

$$x_{\mathrm{rms}} = \sqrt{\lambda_e^{\mathrm{el}} \lambda_e^{\mathrm{ion}}}. \tag{2.18}$$

In our case this is $\sqrt{5(50)} \simeq 15$ cm. Most of the ionization should happen within about 15 cm of the heating coil. This is surprising if we forgot about the random walk, because if the electrons went in straight lines between ionizing collisions they would have quite likely gone all the way to the chamber walls and bounced off before ionizing (or escaped from the plasma in some cases). The ionization would be much more spread out if that happened.

Often we will need to know how long it takes for a particle to diffuse to the chamber wall, diffusion being the name given to the spreading out of particles that

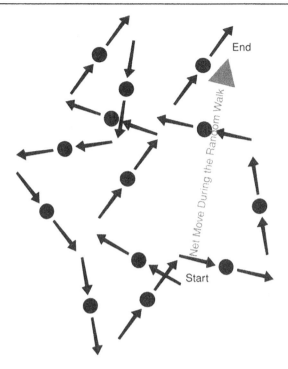

Fig. 2.6. Illustration of a random walk.

takes place in a random walk process. The typical distance traveled in N_s steps is $x_{rms} \simeq \lambda\sqrt{N_s}$. The time to travel a distance λ, which is the collision time, is $\tau_c = \lambda/v$, so N_s steps take a time $t = (\lambda/v)N_s$. Replacing N_s with vt/λ we have

$$x_{rms} \simeq \lambda\sqrt{vt/\lambda} \simeq \sqrt{\lambda vt}. \qquad (2.19)$$

The root mean square distance traveled, x_{rms}, increases as the square root of time, because the mean square distance the particles spread out by increases linearly with time.

The diffusion coefficient is $D \equiv \frac{1}{2}\lambda v$ and this can be substituted here to give

$$x_{rms} \simeq \sqrt{2Dt}. \qquad (2.20)$$

(The numerical coefficient in the square root will usually be quoted as 4 or 6, depending on how many dimensions we allow the random walk to be in.) The speed $v = \lambda/\tau_c$ and so D can be written

$$D = \frac{1}{2}\lambda v = \frac{1}{2}\lambda^2/\tau_c = \frac{1}{2}v^2\tau_c, \qquad (2.21)$$

or, using $\nu \equiv 1/\tau_c$,

$$= \frac{1}{2}\lambda^2\nu = \frac{1}{2}v^2/\nu. \qquad (2.22)$$

As another illustration of these ideas, suppose we wanted to understand the role metastable atoms play in ionization in a discharge containing He or Ar, both of which have long-lived "metastable" states. Metastable atoms, when they collide with each other, can sometimes ionize one of the atoms using the potential energy from the other. Metastables are also lost to the wall and it is useful to know whether most metastables make it to the wall or not before they get involved in an ionization event. We thus need to estimate the time a metastable needs to reach the wall. Suppose we know that at the pressure in our system their mean free path was $\lambda \simeq 10$ cm and their thermal speed was given by $\frac{1}{2}mv_{th}^2 \simeq$ an energy corresponding room temperature (i.e., about $\frac{1}{30}$ of an electron volt). We can find their thermal speed from this, and then the time to diffuse to the wall and the diffusion coefficient D.

An electron volt is roughly 1.6×10^{-19} joules, so our metastable has about $\frac{1}{30}$ of this or about 5×10^{-21} joules. Its mass, if it is He, is $m \simeq 4$ proton masses or about 7×10^{-27} kg. Now $\frac{1}{2}mv_{th}^2 = 5 \times 10^{-21}$ J means $v_{th}^2 \simeq 10^6$ (m/s)^2, and $v_{th} \simeq 10^3$ m/s. The time t to diffuse to the wall from the center of the chamber will be given by Equation (2.19) using $x_{rms} = R$ so that

$$R \simeq \sqrt{\lambda v t} \qquad (2.23)$$

or

$$t = \frac{R^2}{\lambda v}. \qquad (2.24)$$

In our example $\lambda = 0.1$ m. So $t = (0.3)^2/(0.1)(10^3) \simeq 1$ ms. Whether or not this is faster than the time for collision with another metastable depends on the density of metastables.

The mean free paths that were used here must in reality all be calculated as well, from the cross sections and number densities, taking care to use the correct density. It is a very common error, when calculating an electron mean free path, to say $\lambda = \frac{1}{N\sigma}$ and to set N equal to, for instance, the electron density, when the background particle being collided with is the neutral species. The density N to be used is the density of the background species with which our test particle is colliding. In our example of the metastables, their diffusion consists (almost entirely) of a random walk between collisions with ground-state neutrals (not electrons or other metastables). The number density of ground-state neutrals is the N we should use to find the λ that appears in the diffusion coefficient, along with the cross section σ for metastables colliding elastically with ground-state neutrals. In most plasmas used for materials processing the neutral number density greatly exceeds the charged particle density; consequently, the most common collisions of all species are usually with neutrals. Other types of collisions with other particles would have a mean free path λ proportional to one over the number density of the appropriate species.

In this description of random walks we used examples of electrons and neutrals undergoing random walks. Ions were left out because they do not tend to have the same kind of random walk. The random walks we envisage have a changing direction of motion after each step of the walk. Instead, positive ions are usually pulled out of the plasma by the electric field. Collisions slow them down and give them small amounts of velocity "sideways," but their motion is mainly in the direction of the electric field, so it does not much resemble the random walk we described. The motion of the neutrals is nearly random in direction because the neutrals are unaffected by the electric field. The neutrals pick up some small amount of energy from the charged particles, and they typically have a mean energy corresponding to an absolute temperature between room temperature and twice room temperature. The electrons are affected by the electric field. In particular, the electrons are nearly all confined inside the plasma volume by the electric fields, which are very strong near the walls. Unless the electrons are in the strong sheath field near the wall, however, the electric potential variation is often less than the electrons' thermal energy. Inside the plasma the electrons bounce around, colliding with neutrals and the electric fields near the walls. Their motion tends to be random in direction like that of the neutrals but unlike that of the ions. The energy that is given to the plasma by the external power supply is almost always given (at least at first) to the electrons, which are responsible for sustaining the plasma by causing ionization. At least some of the electrons are consequently quite energetic, with energies up to and beyond the ionization threshold, which is typically 10–20 eV. Their average energy is usually a few electron volts. Positive ions acquire high energy by the time they hit the wall. Thus much of the power input to the plasma is carried out again by the ions.

2.6 The Continuity Equation

The next stage in developing a description of particle transport is the "continuity equation," from which the "diffusion equation" may be derived in some circumstances. The continuity equation describes the continuity, or conservation, of particles. Suppose the number density of particles is n. In a volume ΔV there are $n \Delta V$ particles. The number in the volume can change because particles pass through the surface of the volume, or because particles are created or destroyed inside the volume. If the net rate of production of particles per unit volume is S, then the net rate of production in ΔV is $S \Delta V$. The rate of change of the number in ΔV is $\frac{\partial}{\partial t}(n \Delta V)$ and is equal to the net number produced per second in ΔV, $S \Delta V$, minus the number leaving through the surface per second:

$$\frac{\partial}{\partial t}(n \Delta V) = S \Delta V - (\text{number leaving the surface per second}). \quad (2.25)$$

If we divide by ΔV, we have

$$\frac{\partial n}{\partial t} = S - (\text{number leaving the surface per second})/\Delta V. \quad (2.26)$$

The number crossing any surface can be found from the flux per unit area Γ. Γ is equal in magnitude to the number of particles per second passing through a unit area when the normal $\hat{n}$ to that area is parallel to Γ. If the normal is not parallel to the flux then the number per second crossing a unit area is $\Gamma \cdot \hat{n}$. Now in reality Γ will often not be constant all across a unit area, but if the area is a small area Δs then the number crossing it per second is $\Gamma \cdot \hat{n} \, \Delta s$. Finally, for a closed surface we can find the total number going out through it per second by integrating over the surface area,

$$\text{number per second crossing surface} = \oint \mathbf{\Gamma} \cdot \hat{n} \, ds. \tag{2.27}$$

Here $\hat{n}$ is the outward normal. In the equation for $\frac{\partial n}{\partial t}$ we have this quantity divided by ΔV. The integral of a flux over the entire surface of a very small volume, when divided by that volume, is the net amount leaving per unit volume. The net amount leaving per unit volume, which could be described as the flux produced per unit volume, is called the divergence of the flux of that quantity. In other words, when ΔV is very small (strictly when $\Delta V \to 0$)

$$(\text{number leaving through surface per second})/\Delta V = \oint \mathbf{\Gamma} \cdot \hat{n} \, ds / \Delta V = \nabla \cdot \mathbf{\Gamma} \tag{2.28}$$

is the definition of $\nabla \cdot \mathbf{\Gamma}$. Our conservation equation is then

$$\frac{\partial n}{\partial t} = S - \nabla \cdot \mathbf{\Gamma}. \tag{2.29}$$

This is an exact equation for the number density n. Approximate expressions for $\mathbf{\Gamma}$ will be described next.

2.7 The Particle Flux

The number of charged particles crossing a unit area per second, $\mathbf{\Gamma}$, is usually approximated as

$$\mathbf{\Gamma} = -D\nabla n + n\mu\mathbf{E}, \tag{2.30}$$

where D is the diffusion coefficient, μ is the mobility, $\mathbf{E}$ is the electric field, and n is the particle number density. The term involving the electric field will be described first.

The second term, $n\mu\mathbf{E}$, is included because charged particles are often expected to have a drift velocity $\mathbf{v}_d$ proportional to the electric field, with the "mobility" μ being the constant of proportionality,

$$\mathbf{v}_d = \mu\mathbf{E}. \tag{2.31}$$

The mobility μ will (usually) be negative if the particles are negative.

This drift velocity $\mathbf{v}_d$ means that in one second particles move on average $\mathbf{v}_d$ meters. If we consider a unit area at right angles to $\mathbf{v}_d$ then all the particles in a volume of length $\mathbf{v}_d$ and unit area, that is, a volume of v_d cubic meters "upstream" of the area, flow through the area in a second. The number of particles in this volume is $nv_d = n\mu E$, and hence this is the magnitude of the flux per unit area created by the drift.

The diffusion term is caused by the random motion of particles. If there is a large number of particles at one point A and a small number at another point B nearby, then if their motion is isotropic more will be leaving A going toward B than will be leaving B going toward A. The net flux is from the high-density point to the low-density point. Since ∇n points in the direction of increasing density, this diffusive flux is proportional to $-\nabla n$. The constant of proportionality is the diffusion coefficient D, hence the diffusive part of the flux is $-D\nabla n$.

These two contributions to the flux are approximations based on experiment. The expressions given are expected to be accurate enough to be useful when the mean free path is very small compared to the system size.

In equilibrium there is no flux, by definition of equilibrium. In one dimension x and using $E = -\frac{d\Phi}{dx}$, where Φ is the electrostatic potential, the flux is

$$\Gamma = -D\frac{dn}{dx} - n\mu\frac{d\Phi}{dx}. \tag{2.32}$$

This can be solved by an "integrating factor" method in many cases. When $\Gamma = 0$, the solution for n (which can be checked by substitution) is

$$n = n_o \exp\left(-\frac{\mu}{D}\Phi\right). \tag{2.33}$$

In equilibrium, however, we expect the Boltzmann relation

$$n = n_o \exp(-q\Phi/k_B T) \tag{2.34}$$

to hold, where T is the particle temperature and q is the particle charge. These equations are consistent provided that

$$\frac{\mu}{D} = q/k_B T, \tag{2.35}$$

which is known as the Einstein relation between μ and D. (D is positive; if q is negative, μ is also negative.)

One of the useful consequences of these equations is that if we know the density of electrons n and if they have a temperature T we can take the natural logarithm of the Boltzmann relation to find the electrostatic potential from n. Since for electrons $q = -e$,

$$\Phi = \frac{k_B T}{e}\ln\left(\frac{n}{n_o}\right). \tag{2.36}$$

This expression implies that Φ is zero at the point where n is equal to n_o, which is often chosen to be the maximum density. The equation is not likely to work for the ions since they are not close to equilibrium.

In the expression for the electron flux the two contributions to the flux are much larger than the corresponding terms in the ion flux. However, the overall electron flux is often equal to the overall ion flux. The explanation is that the two very large terms, $-D_e\nabla n_e$, and $n_e\mu_e E$, in the electron flux usually nearly cancel but the difference is still large enough to be equal to the ion flux. By contrast, in the expression for the ion flux the small fluxes $-D_i\nabla n_i$ and $n_i\mu_i E$ are usually in the same direction. Because the two contributions to the electron flux nearly cancel, the electrons can be close to being in equilibrium, even though the ions are not close to equilibrium.

2.8 Feedback in Plasmas

For any method of supplying power to a plasma, there is a limit to the plasma density that can be sustained. Increasing density eventually decreases the heating efficiency, but the way this comes about may be very different from one case to another. To understand a plasma reactor it is usually necessary to deduce what is the mechanism that limits the plasma density. Two examples will now be given, in the first of which the feedback is largely due to the external circuit whereas in the second the feedback is done by the plasma itself.

In a dc discharge plasma where a dc voltage is applied to the ends of a long thin cylindrical plasma, a situation can be created where most of the plasma is what is known as a "positive column." There is an electric field of magnitude E_z along the z axis of the cylinder, due to the applied voltage. The field E_z is roughly equal in magnitude to the applied voltage divided by the length L of the cylinder. The field is responsible for heating the electrons and hence for the ionization. The loss of the plasma is primarily radial, to the walls of the cylinder. In this plasma as described so far, a doubling of the plasma density does not directly affect the rate at which an electron is heated, and it may not directly affect the time for a particle to be lost from the plasma, either. Doubling the plasma density will however double the plasma current, if E_z is fixed. In reality the external circuit has a large "ballast" resistor in it, which is in series with the voltage source, and when the current doubles the voltage dropped across the ballast resistor doubles, decreasing the voltage applied to the plasma and hence decreasing E_z. If E_z decreases, the heating rate per electron decreases, lowering the ionization rate per electron and hence the density and providing at least some of the negative feedback needed for stability. The ballast resistor in the external circuit is (partially) responsible for the feedback in this case.

In contrast in an inductively coupled plasma increasing the plasma density decreases the skin depth to which the electromagnetic fields penetrate the plasma. If the fields penetrate less far the heating rate will eventually go down and limit the density that can be achieved. In this case the plasma itself provides the negative feedback, although additional negative feedback caused by the external circuit is also likely.

Exercises

1. Neutral gas is inlet to a cylindrical chamber of radius $a = 30$ cm and height 20 cm through an inlet 10 cm above the base of the cylinder. The outlet is opposite the inlet. If the mean free path for neutral particles is $\lambda_n = 1$ cm, how many steps of the random walk does it typically take for a particle to go 10 cm (from the inlet to the cylinder base), 20 cm (from the base to the top), and 60 cm (from inlet to outlet)?

   How fast would the gas have to flow from inlet to outlet, so that particles flowed to the outlet in about the same time they take to reach the base? The neutral velocity is v_n. Express the flow velocity in terms of v_n. Comment on the uniformity of etching, if a small percentage of these neutral particles are the etchants, which stick to the base with a probability of one.

2. Gas is pumped into a cylindrical chamber of radius $a = 30$ cm and height $h = 20$ cm at a rate of N_{in} particles per second. If the gas input rate is specified to be 20 sccm (standard cubic centimeters per minute) what is N_{in}? Gas is

pumped out of the chamber through an opening of area A_{out}. Assume that the density in the chamber is uniform and that the gas particles' motion is isotropic. If every particle striking the area A_{out} is lost (i.e., pumped out), what will the gas number density in the chamber be? (See Chapter 6, if you are not familiar with solving diffusion and transport equations.)

Now treat the transport of gas in the chamber as being a one-dimensional flow, with the flux being purely diffusive from the inlet to the outlet. The mean free path of neutrals is $\lambda_n \ll a$, and the distance from inlet to outlet is $2a$, with the inlet at $z = 0$ and the outlet at $z = 2a$. All the particles that reach the outlet are lost; what should the boundary condition at $z = 2a$ be? What is the solution of the diffusion equation, in terms of $n_0 \equiv n(z = 0)$?

Using the fact that N_{in} particles flow in, per second, into an area of A_{cyl}, find n_0, by setting the diffusion coefficient of neutrals $D_n = \frac{1}{2}\lambda_n v_n$, with v_n being the thermal speed of the neutrals. Why is it unrealistic to set D_n equal to a constant? Set $\lambda_n = \frac{1}{\bar{n}\sigma}$, where $\bar{n}$ is the mean neutral density. Can you now find n_0 in terms of σ?

The mean free path cannot be greater than $2a$, or else the model breaks down. Suppose an alternative expression for λ_n is suggested, to allow for this, given by

$$\frac{1}{\lambda_n^{alt}} = \bar{n}\sigma + \frac{1}{2a}. \tag{2.37}$$

Is this a reasonable model? Can you find n_0 using this model?

As an alternative to this approach, suppose that we notice that the flux per unit area is actually

$$\Gamma = -D_n \frac{dn}{dz} + n v_{dr}. \tag{2.38}$$

v_{dr} could be as large as v_n, if there were no collisions. If v_{dr} is very large, we can ignore the diffusion term, and setting $v_{dr} = v_n$, the flux per unit area is $N_{in}/A_{cyl} = n v_n$, so $n = N_{in}/A_{cyl} v_n$, which is constant in space. Compare this value to the minimum density, which is obtained from λ_n^{alt}, given above.

Now suppose that only a fraction $f_{out} \ll 1$ of the wall at $z = 2a$ is open. We shall assume that a fraction f_{out} of the flux of particles striking $z = 2a$ leave and that $(1 - f_{out})$ of the flux is reflected. The *net* flux leaving per unit area (if $v_{dr} = 0$) is

$$\Gamma_{net} = -D_n \frac{dn}{dz}. \tag{2.39}$$

The flux striking $z = 2a$ is $\Gamma_{st} = \frac{1}{4}n v_n$, so $\Gamma_{net} = f_{out}\Gamma_{st}$ gives a new boundary condition:

$$-D_n \frac{dn}{dz} = \frac{1}{4} f_{out} n v_n. \tag{2.40}$$

Assume again that $D_n = \frac{1}{2}\lambda_n v_n$ and $\lambda_n = \frac{1}{\bar{n}\sigma}$. Then $D_n = \frac{v_n}{2\bar{n}\sigma}$ and so

$$-D_n \frac{dn}{dz} = \frac{v_n}{2\bar{n}\sigma} \frac{\Delta n}{2a}, \tag{2.41}$$

where $\Delta n = n_{\max} - n_{\min}$. This is equal to

$$\frac{f_{\text{out}}}{4} v_n n_{\min}. \tag{2.42}$$

Thus

$$\frac{\Delta n}{\bar{n} \sigma a} = f_{\text{out}} n_{\min}, \tag{2.43}$$

where $\bar{n} = \frac{1}{2}(n_{\max} + n_{\min})$. If Δn is small compared to $\bar{n}$, which must be true for our expression for λ_n to be useful, then $\Delta n \simeq f_{\text{out}} \bar{n}^2 \sigma a$.
The flux at the inlet is $\frac{N_{\text{in}}}{A_{\text{cyl}}}$. If we set this equal to

$$-D_n \frac{dn}{dz} = D_n \frac{\Delta n}{2a}, \tag{2.44}$$

we get

$$\frac{N_{\text{in}}}{A_{\text{cyl}}} = \frac{v_n}{2\bar{n}\sigma} \frac{f_{\text{out}} \bar{n}^2 \sigma}{2} = \frac{f_{\text{out}} \bar{n}^2 v_n}{4}. \tag{2.45}$$

This is equivalent to the other expression for the flux, which was used in the boundary condition.

If $\frac{N_{\text{in}}}{A_{\text{cyl}}} = 10^{22}$ m^{-2} s^{-1}, $v_n = 10^2$ ms^{-1}, and $f_{\text{out}} = 0.1$, what is $\bar{n}$? Estimate λ_n and Δn, if $\sigma = 10^{-20}$ m^2 and $a = 0.25$ m.

3. In one dimension, x, the particle flux per unit area is given by $\Gamma_x = n\mu E_x - D \frac{dn}{dx}$. If Γ_x is constant, derive an expression for n in terms of $n(x = 0)$, if necessary, and of Γ_x, first for the simple cases (1) $\mu \neq 0$ but $D = 0$; (2) $D \neq 0$ but $\mu = 0$; and for the more complex case (3) $\mu \neq 0$, $D \neq 0$. In case (3), use an integrating factor method.

4. In one dimension, x, there is a fixed region of positive charge density, with charge per unit volume $\rho = eN_c$, from $x = 0$ to $x = a$. A region of negative charge density can move in the x direction. The negative charge density is $\rho = -eN_c$ and extends from $x = \delta_x$ to $x = a + \delta_x$. The mass per unit volume of the negative charge is mN_c. Sketch the electric field versus x, when $\delta_x > 0$ but $\delta_x < a$. If $\delta_x \ll a$, the average electric force on each negative charge is approximately $\alpha \delta_x$. What is α?

    There is a damping force on each negative charge, equal to $-vm\frac{d(\delta_x)}{dt}$. Write the equation of motion for a negative charge, and solve it in terms of $\delta_x(t = 0)$.

5. The analogy was suggested that the electrons in the plasma behave somewhat like sheep in a field. A sheep might weigh 20 kg and run at 5 ms^{-1}. How heavy would an oxygen atom be, if an electron weighed 20 kg and the oxygen atom was as many times as heavy as the electron as it is in reality? Suggest a suitable analogy for the atom, which has roughly this mass. How fast would an oxygen ion with this mass move, if it had the same energy as the "electron," and if the electron moved at 5 ms^{-1}? How fast would an oxygen atom move, with an energy that is T_n/T_e times lower? Here T_n is the neutral temperature and T_e is the electron temperature. The ratio T_n/T_e is typically of the order 0.02.

3

Plasma Electromagnetics and Circuit Models

3.1 Electromagnetic and Circuit Models

A great deal of insight into the behavior of the plasma can be obtained using simplified models, provided we understand when these models fail. The simplest such type of model is a circuit model, where we represent the plasma and the system enclosing it by passive circuit elements (resistors, capacitors, and inductors) and sometimes diodes. Because the plasma presents different responses depending on the set of "terminals" at which we choose to measure the impedance, we shall need several different circuit models. We also give a brief discussion of electromagnetics, which is important in its own right and also to develop the circuit models.

We begin by considering "equivalent" circuits for systems where the power supply operates at radio frequencies (rf), including ICPs and capacitive discharges. Most processing systems are driven by a radio frequency power supply, at the standard frequency $f = 13.56$ MHz. Some employ microwaves, at 2.8 GHz, and heat the electrons by creating Electron Cyclotron Resonance (ECR). Circuit description of ECR reactors is described next. Direct current (dc) processing systems are rare, although in some regards ECR systems behave as if they are dc. Application of a circuit model to find the voltage across an insulating surface-layer is the last topic in this chapter.

To make progress, including in developing circuit models, we now review some results from electromagnetics.

3.2 Plasma Electromagnetics

In several of the important plasma processing reactors that are in use, there are rf or microwave electromagnetic fields that play a crucial role in sustaining the plasma. A detailed description of these fields is an enormously complicated undertaking. The analytic results available provide clues as to what happens in reality, but analytic methods are not adequate in general.

The description of the electromagnetic fields in an ICP is a good example to consider, for several reasons. At present, the ICP is one of the most used types of reactor. It is also an example where we can hope to get a meaningful answer from a simple analysis. Conventionally, we do not solve the full set of Maxwell's equations in order to describe the ICP fields. The ICP operates at radio frequencies of about $f \sim 10^7$ s^{-1}, and so the wavelength $\lambda = c/f$ is $\lambda \simeq 30$ m. This is much larger than the ICP dimension a, which means that the fields inside the ICP can be considered "static" in a certain sense. There is very little change in the phase of a wave, over the distance a, when a is only 1% of the wavelength. In Maxwell's equations, the "low" frequency means that even in the absence of the plasma, the displacement current density, $\frac{\partial \mathbf{D}}{\partial t}$, is not very important in Ampère's law,

$$\nabla \times \mathbf{H} = \mathbf{J} + \frac{\partial \mathbf{D}}{\partial t}. \tag{3.1}$$

In the presence of a plasma, with $\mathbf{J} = \sigma \mathbf{E}$ and σ large, this may be even more likely to be true.

To see this, first consider the vacuum fields inside the tank and Ampère's law. Since the current density $\mathbf{J}$ is zero in the vacuum,

$$\nabla \times \mathbf{H} = \frac{\partial \mathbf{D}}{\partial t}. \tag{3.2}$$

The vacuum wavelength is λ. The displacement current is probably big enough to cause $\nabla \times \mathbf{H}$ to be of the order H/λ, since the derivative of H with respect to position in the vacuum fields in free space should be about this big. But the components of $\nabla \times \mathbf{H}$ are much bigger than this, because the fields must fit into the tank; $\frac{\partial H_r}{\partial z} \sim \frac{\partial H_z}{\partial r} \sim \frac{H}{a}$, where a is the vacuum tank dimension. Since $\lambda \gg a$, we can neglect the displacement current.

In contrast, in the plasma $\mathbf{J} = \sigma \mathbf{E}$, and σ is typically large. The displacement current $\frac{\partial \mathbf{D}}{\partial t} \sim \omega \epsilon \mathbf{E}$, for a sinusoidal wave, and $\sigma \gg \omega \epsilon$. Thus the conduction current may well not be negligible – and in practice it is important.

Now if $\nabla \times \mathbf{H} = \sigma \mathbf{E}$, and since $\nabla \times \mathbf{E} = -\frac{\partial \mathbf{B}}{\partial t}$, then

$$\nabla \times (\nabla \times \mathbf{E}) = -\mu \frac{\partial}{\partial t}(\nabla \times \mathbf{H}) = -\mu \frac{\partial}{\partial t}(\sigma \mathbf{E}). \tag{3.3}$$

If σ is constant,

$$\nabla \times (\nabla \times \mathbf{E}) = -\mu\sigma \frac{\partial \mathbf{E}}{\partial t}. \tag{3.4}$$

Now $\nabla \times (\nabla \times \mathbf{E}) = \nabla(\nabla \cdot \mathbf{E}) - \nabla^2 \mathbf{E}$, and if the free charge density ρ is zero (or is not contributing to the electric field we are interested in, perhaps because it does not oscillate at the right frequency) then $\nabla \cdot \mathbf{E} = 0$ and $\nabla \times (\nabla \times \mathbf{E}) = -\nabla^2 \mathbf{E}$.

It is important to remember that the right-hand side is the Laplacian acting on a vector. The term $\nabla^2(E_\phi \hat{\phi})$ is not equal to $\hat{\phi}\nabla^2 E_\phi$, for example; see Ref. [57]. To clarify this point, suppose ϕ is defined so that the unit vector $\hat{\phi}$ points in the direction

$$\hat{\phi} = \hat{x} \sin\phi - \hat{y}\cos\phi. \tag{3.5}$$

Since

$$\sin\phi = \frac{y}{\sqrt{x^2 + y^2}} \quad \text{and} \quad \cos\phi = \frac{x}{\sqrt{x^2 + y^2}}$$

we have

$$\hat{\phi} = (\hat{x}y - \hat{y}x)/\sqrt{x^2 + y^2}. \tag{3.6}$$

Now

$$\nabla^2 \hat{\phi} = \frac{\partial^2 \hat{\phi}}{\partial x^2} + \frac{\partial^2 \hat{\phi}}{\partial y^2}.$$

Taking the first term, we get

$$\frac{\partial^2 \hat{\phi}}{\partial x^2} = \hat{x} y \frac{\partial^2}{\partial x^2} \left[\frac{1}{\sqrt{x^2 + y^2}} \right] - \hat{y} \frac{\partial^2}{\partial x^2} \left[\frac{x}{\sqrt{x^2 + y^2}} \right]$$

$$= (x^2 + y^2)^{-5/2} [\hat{x} y (2x^2 - y^2) + \hat{y} 3y^2 x].$$

By analogy, the other term must be

$$\frac{\partial^2 \hat{\phi}}{\partial y^2} = \frac{-\hat{y} x (2y^2 - x^2) - \hat{x} 3x^2 y}{(x^2 + y^2)^{5/2}}. \tag{3.7}$$

Combining these, we obtain

$$\frac{\partial^2 \hat{\phi}}{\partial x^2} + \frac{\partial^2 \hat{\phi}}{\partial y^2} = (x^2 + y^2)^{-5/2} [\hat{x}(2x^2 y - y^3 - 3x^2 y) + \hat{y}(3y^2 x - x2y^2 + x^3)]$$

$$= (x^2 + y^2)^{-3/2} [x\hat{y} - y\hat{x}]$$

$$= -\hat{\phi}/(x^2 + y^2) = -\frac{\hat{\phi}}{r^2}.$$

Then

$$\nabla^2 (\hat{\phi} E_\phi) = \hat{\phi} \left[\nabla^2 E_\phi - \frac{E_\phi}{r^2} \right]. \tag{3.8}$$

In an accurate calculation, the second term is important. In the analytic estimates that follow, it will be neglected.

3.2.1 Electromagnetic Fields in an ICP

From Equation (3.4) we can obtain the result

$$\nabla^2 E_\phi = \mu \sigma \frac{\partial E_\phi}{\partial t}. \tag{3.9}$$

Thus if, in the region $z < 0$ below the coil, E_ϕ varies as $\exp[j(\omega t + \kappa z)]$, then $\kappa^2 = -j\omega\mu\sigma$. The solution for κ is $\kappa = (1 - j)/\delta$, with $\delta = (\omega\mu\sigma/2)^{-1/2}$ and

$$E_\phi = E_0 \exp \left[j \left(\omega t + (1 - j)\frac{z}{\delta} \right) \right]$$

$$= E_0 \exp \left(\frac{z}{\delta} \right) \exp \left[j \left(\omega t + \frac{z}{\delta} \right) \right].$$

Using Faraday's law, $\nabla \times \mathbf{E} = -\frac{\partial \mathbf{B}}{\partial t}$, the component in the radial direction is

$$-\frac{\partial E_\phi}{\partial z} = -\frac{\partial B_r}{\partial t}. \tag{3.10}$$

This implies that

$$\frac{(1+j)}{\delta} E_0 = j\omega B_0, \tag{3.11}$$

so that $B_0 = (1 - j)E_0/(\delta\omega)$ and

$$B_r = (1 - j)\frac{E_0}{\delta\omega} \exp\left(\frac{z}{\delta}\right) \exp\left[j\left(\omega t + \frac{z}{\delta}\right)\right]. \tag{3.12}$$

The phase difference between E and B is $\pi/4$, as is always the case in a "good conductor" with a real conductivity. (The plasma conductivity may be nearly real, at high pressure, but it is complex at lower pressure.) The impedance presented by such a medium has equal real and imaginary parts. This has immediate consequences for the impedance presented by the plasma to the external circuit.

In the remainder of this section, we shall use these expressions to analyze a situation where an "antenna" with a fixed current is in contact with a plasma. In the next section we shall use a simpler approach based on the stored energy and power deposition in the plasma, to extend the analysis here, to deduce how the impedance changes when we introduce a nonconducting gap between antenna and plasma.

These results for E_ϕ and B_r can be related directly to the input impedance of the antenna. The spiral nature of the antenna will not be included, nor will the variation of the fields with radius r. Instead, suppose that the antenna consists of N_a turns, each of circumference $2\pi\bar{r}$, where $\bar{r}$ is the mean radius of the antenna. The (phasor) antenna voltage is then $V_a = 2\pi\bar{r}N_a E_\phi(z = 0) = 2\pi\bar{r}N_a E_0 e^{j\omega t}$.

The (phasor) antenna current I_a is the total (phasor) current on the "current sheet" between $r = 0$ and $r = R_m$, divided by N_a. (Since $V_a \sim N_a$ and $I_a \sim 1/N_a$, the power varies as $V_a I_a$, which is independent of N_a.) The (phasor) current per unit length J_s (the current being in the ϕ direction, but the length being in the radial direction) is just equal to the radial component of $H = B/\mu$, according to the usual boundary condition. The total current is this multiplied by the maximum radius R_m, which is $R_m B_0/\mu = I_a N_a$, and so $I_a = \frac{R_m B_0}{N_a \mu}$.

The impedance seen at the input to the antenna is thus

$$Z_a = \frac{V_a}{I_a} = \frac{2\pi\bar{r}N_a E_0}{(R_m B_0/N_a\mu)} = 2\pi(\bar{r}/R_m)\mu N_a^2 \frac{E_0}{B_0}, \tag{3.13}$$

but we found above that

$$E_0/B_0 = \frac{\delta\omega}{1 - j} = \left(\frac{1+j}{2}\right)\delta\omega \tag{3.14}$$

so if $\frac{\bar{r}}{R_m} = \frac{1}{2}$, then

$$Z_a = \frac{\pi}{2}\mu\delta\omega N_a^2(1 + j). \tag{3.15}$$

This seems to imply that the plasma creates an impedance in the primary circuit consisting of a resistance

$$R_a = \frac{\pi}{2}\mu\delta\omega N_a^2 \tag{3.16}$$

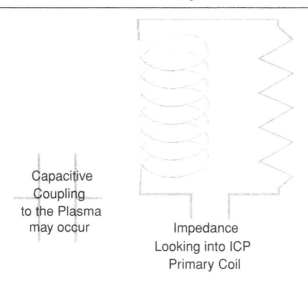

Capacitive
Coupling
to the Plasma
may occur

Impedance
Looking into ICP
Primary Coil

Fig. 3.1. Equivalent circuit for an inductive discharge.

and an inductance

$$L_a = \frac{\pi}{2}\mu\delta N_a^2. \tag{3.17}$$

This is shown schematically in Fig. 3.1.

The values of these components will be altered if the conductivity is complex. The conductivity is real in a static field; we now briefly indicate how an rf field affects the conductivity. Suppose an rf electric field is present in the plasma so that the equation of motion of the electron,

$$m\frac{d\mathbf{v}}{dt} = -e\mathbf{E} - m\nu\mathbf{v}, \tag{3.18}$$

has two force terms on the right, an electric force and a frictional drag force. If we assume the velocity varies sinusoidally we can replace it with a phasor velocity. If the angular frequency of the rf is ω, the phasor equation becomes

$$j\omega m\mathbf{v} = -e\mathbf{E} - m\nu\mathbf{v}. \tag{3.19}$$

Then the velocity is

$$\mathbf{v} = -\frac{e\mathbf{E}}{m(\nu + j\omega)}. \tag{3.20}$$

The rf field introduced an extra factor of $(1 + j\omega/\nu)$. This modifies the conductivity by the same factor. It will also affect the impedance that the plasma presents to the primary circuit (see the exercises).

Electromagnetic Stored Energy and Power Deposition in an ICP

In this section, armed with the result from the previous section, we continue discussing how to use our electromagnetic results to set up a circuit model to investigate, among other things, the feedback in the ICP. The first circuit we are led to consider in the ICP includes the primary coil, which powers the plasma, and the secondary, which is the plasma itself. If we treat the plasma as being a good conductor with a constant conductivity (but see Refs. [58], [59]) we should be able to investigate how the power deposited in the plasma varies with conductivity, by estimating the inductance and resistance and using them in the circuit model. The values of the inductance and resistance will be recalculated, allowing for an insulating gap between the coil and the plasma. The plasma conductivity is not nearly constant in space or time. For now we just assume it is proportional to the density so that increasing density controls the power deposition through the conductivity.

We now repeat the earlier calculation, in a different order and using the "time-domain" instead of phasors. (The results of the two methods should be compared.)

If the primary coil is thought of as a current sheet it will have a surface current of J_s A/m flowing along the surface. If the coil carries a total current $I_T = I_a N_a$ and this flows along the surface perpendicularly to the radial direction, between the radii $r = 0$ and $r = R_m$, then if the current density is uniform in radius the current per unit radius is $J_s = I_a N_a / R_m$.

If the coil is a perfect conductor there is no magnetic field inside it, and the boundary condition on H states that H next to the coil is equal to J_s but points in the radial direction. If the coil is not a perfect conductor the current is carried in a layer, which may correspond to the entire coil thickness. The magnetic field inside the coil goes to zero behind the current layer. H increases to J_s at the bottom surface of the coil. The H field in the plasma decays in the vertical direction, with a decay length equal to the skin depth δ. If $z = 0$ by the coil and z decreases as we go down, then we expect

$$\mathbf{H} = \mathbf{a}_r I_a N_a / R_m \, e^{+z/\delta} \cos(\omega t + z/\delta), \tag{3.21}$$

where I_a is the current in the antenna. Faraday's law,

$$\nabla \times \mathbf{E} = -\frac{\partial \mathbf{B}}{\partial t}, \tag{3.22}$$

can be used to find $\mathbf{E}$. When using Maxwell's equations to find one field from another, we first note that the divergence equations each only contain one of the fields so they cannot be used for this purpose. We also have a choice of curl equations, each of which has the curl of a field set equal to the time derivative of the other field. (The other curl equation is Ampère's law.) It is considerably easier in general to use the equation that takes the curl of the field we already know, and integrate that curl over time to find the other field. In the case we are considering here this would still be a little easier, but since the fields only depend on z and t, the curl of $\mathbf{E}$ is given by

$$\nabla \times \mathbf{E} = \mathbf{a}_r \left(-\frac{\partial E_\phi}{\partial z} \right) + \mathbf{a}_\phi \left(\frac{\partial E_r}{\partial z} \right). \tag{3.23}$$

Because $\frac{\partial \mathbf{B}}{\partial t}$ has only a radial component, $\partial E_r / \partial z = 0$ and

$$-\frac{\partial E_\phi}{\partial z} = \frac{I_a N_a}{R_m} \omega \mu e^{z/\delta} \sin(\omega t + z/\delta). \tag{3.24}$$

Therefore

$$E_\phi = \frac{I_a N_a \omega \mu \delta}{2R_m} e^{z/\delta} (\cos(\omega t + z/\delta) - \sin(\omega t + z/\delta)). \tag{3.25}$$

Power Deposition

The power deposited can be found from E_ϕ, again assuming σ is constant, and also that it is real (which is not true in low-pressure ICPs – see Refs. [58], [59]), since the current per unit area $\mathbf{J}$ is $\sigma \mathbf{E}$ and the heating rate per unit volume is

$$P = \mathbf{J} \cdot \mathbf{E} = \sigma E^2. \tag{3.26}$$

Then the time-averaged heating rate per unit volume is

$$\bar{P} = \frac{1}{2}\sigma E_o^2 e^{2z/\delta}, \tag{3.27}$$

where $E_o = I_a N_a \omega \mu \delta/(\sqrt{2}R_m)$. Integrating this over z gives the heating rate per unit area of the coil,

$$\int_{-\infty}^{0} \bar{P}\,dz = \frac{1}{2}\sigma E_o^2 \frac{\delta}{2} e^{2z/\delta} \Big|_{-\infty}^{o} = \sigma E_o^2 \delta/4. \tag{3.28}$$

The total heating rate is this times πR_m^2, or

$$P_{\text{tot}} = \sigma \left(\frac{I_a N_a \omega \mu \delta}{\sqrt{2}R_m}\right)^2 \frac{\delta}{4} \pi R_m^2 \tag{3.29}$$

$$= \sigma (I_a N_a \omega \mu)^2 \pi (\delta/2)^3.$$

But since $\delta = (\frac{\omega \mu \sigma}{2})^{-\frac{1}{2}}$, where μ is the permeability,

$$P_{\text{tot}} = \sigma (I_a N_a \omega \mu)^2 \frac{\pi}{8} \left(\frac{2}{\omega \mu \sigma}\right)^{3/2} = \frac{\pi (I_a N_a)^2}{2\sqrt{2}} \left(\frac{\mu \omega}{\sigma}\right)^{1/2}. \tag{3.30}$$

This implies that increasing the conductivity decreases the power. As the plasma becomes more like a perfect conductor, the shielding gets better and the skin depth gets smaller. The plasma is able in this way to limit the power deposited as the density increases.

Stored Energy

The energy stored in an inductor is $W_L = \frac{1}{2}Li^2$, where L is the inductance and i the current. However, the energy stored per unit volume in a magnetic field is

$$w_m = \frac{1}{2}\mathbf{B} \cdot \mathbf{H}. \tag{3.31}$$

The integral of w_m over volume is equal to W_L. The inductance L can be found by setting $L = \frac{2}{i^2}$ times the volume integral of w_m. The total energy stored in the magnetic field is calculated similarly to the power deposition. First we average over time and then integrate over space to get

$$W_L = \frac{\pi}{2} R_m^2 \int_{-\infty}^{o} \frac{\mu}{2} \left(\frac{I_a N_a}{R_m} \right)^2 e^{2z/\delta} dz \tag{3.32}$$

$$= \frac{\pi}{8} \mu (I_a N_a)^2 \delta.$$

Now $W_L = \frac{1}{2} L i^2$, where i is the instantaneous current, but the maximum current flowing in the antenna is $I_a = I_T / N_a$, where N_a is the number of turns, and so the time average of W_L is $\frac{1}{4} L I_a^2$. The inductance is thus estimated to be $L_p = \frac{\pi}{2} \mu \delta N_a^2$. Similarly from the point of view of the primary circuit the time-average power deposited has the form $P_{tot} = \frac{1}{2} I_a^2 R_p = \frac{1}{2} (I_T / N_a)^2 R_p$, where

$$R_p = \frac{\pi}{\sqrt{2}} \left(\frac{\mu \omega}{\sigma} \right)^{\frac{1}{2}} N_a^2 = \frac{\pi}{2} \omega \mu \sqrt{\frac{2}{\omega \mu \sigma}} N_a^2 = \frac{\pi}{2} \omega \mu \delta N_a^2. \tag{3.33}$$

These confirm the results obtained earlier. Now we extend this analysis to a case with an insulating region between the antenna and the plasma.

Stored Energy in an Insulating Gap

Outside the plasma, the fields are nearly "static." In a gap between the current sheet and the plasma, $H_r = J_s$; there is no E_ϕ if the field is static. At the plasma surface, both E_ϕ and H_r should be continuous. However, we shall impose continuity of H_r, we find E_ϕ in the plasma from H_r, and we assume (again) that E_ϕ in the vacuum is small.

The stored energy in the vacuum region, W_{vac}, which is of height z_p, is found from the stored energy per unit volume in the vacuum, $w_{vac} = \frac{1}{2} B_r H_r$, to be

$$w_{vac} \pi R_m^2 z_p = \frac{1}{2} (B_r H_r) \pi R_m^2 z_p \tag{3.34}$$

$$= \frac{1}{2} \mu_o J_s^2 \pi R_m^2 z_p.$$

If there are N_a coils in the radial distance R_m, then $J_s = I_a N_a / R_m$. Thus

$$W_{vac} \simeq w_{vac} \pi R_m^2 z_p \simeq \frac{1}{2} \mu_o (I_a N_a / R_m)^2 \pi R_m^2 z_p \tag{3.35}$$

so that

$$W_{vac} \simeq \frac{\pi}{2} \mu_o I_a^2 N_a^2 z_p. \tag{3.36}$$

From this we deduce

$$L_{vac} = \frac{\pi}{2} \mu_o N_a^2 z_p. \tag{3.37}$$

The total inductance seen in the primary is therefore $L_{pr} = L_{vac} + L_p = \frac{\pi}{2} \mu_o N_a^2 (z_p + \delta)$, with impedance $j\omega L_{pr}$. The resistance seen in the primary is $R_p = \frac{\pi}{2} \omega \mu_o N_a^2 \delta$. Then $R_p / (\omega L_{pr}) = \delta / (z_p + \delta)$.

3.3 Circuit Model of an ICP – Negative Feedback in the Primary Circuit

We can use this information to investigate possible sources of negative feedback in the ICP. The question addressed is: as the plasma density increases, how is the power supply to the plasma decreased, so that an equilibrium is reached? We begin with the external circuit. The primary circuit power supply typically resembles a current source, with current I_s and with a large resistor R_s in parallel with it. If the plasma (plus insulating gap) resembles an inductor L_{pr} and a resistor R_p in series then the plasma presents an impedance

$$Z_p = R_p + j\omega L_{pr} \tag{3.38}$$

and the current in the primary is equal in amplitude to

$$I_a = \frac{I_s R_s}{\sqrt{(R_s + R_p)^2 + (\omega L_{pr})^2}}. \tag{3.39}$$

If R_s is very large compared to R_p and ωL_{pr} then the feedback from the plasma has little effect on the current I_a and in reality this is usually the case. In fact as the conductivity increases both R_p and L_{pr} decrease and so this circuit will provide a roughly constant current of I_a at high plasma densities.

The plasma itself limits the maximum density since, as we have just shown, the absorbed power is $\frac{1}{2}I_a^2 R_p$ and this decreases as the skin depth decreases.

The behavior of the external circuit would limit the absorbed power as the skin depth increased, if R_p or ωL_{pr} became large enough. The time-averaged absorbed power is

$$\bar{P}_{tot} = \frac{1}{2}\frac{I_s^2 R_s^2 R_p}{(R_s + R_p)^2 + (\omega L_{pr})^2}. \tag{3.40}$$

According to our estimates $R_p = \omega L_p = \alpha\delta$, where $\alpha \equiv \frac{\omega\pi}{2}\mu N_a^2 \simeq 2\pi(1.3 \times 10^7)\frac{\pi}{2} \cdot 4\pi \times 10^{-7} \times 3^2 \simeq 1{,}500$. The maximum $\bar{P}_{tot}$ as a function of σ should occur when $d\bar{P}_{tot}/d\delta = 0$, which is when $R_s^2 = \alpha^2(2\delta^2 - z_p^2)$. This would imply a very large skin depth $\delta \gg L$, the system height. Several approximations we made break down if $\delta \geq L$ however.

In the rest of this section, we redefine δ to be equal to the skin depth δ plus z_p, to simplify the notation. One reason why δ cannot become large is that even in the absence of a plasma the fields decay with z with a skin depth δ of a few centimeters, owing largely to the finite coils and the conducting vacuum tank. We can model the upper limit on the skin depth by setting the actual skin depth to be $\delta\delta_{max}/(\delta + \delta_{max})$, where δ is the skin depth in an infinite plasma. The resistance will be modified by this change in the spatial scale of the fields, and it is the resistance that is important for the absorbed power.

In dealing with the resistance R_{tot} we have to go back to a point in the derivation where we can distinguish between the dependence on σ and the dependence on the effective skin depth $\hat{\delta} = \frac{\delta\delta_{max}}{\delta + \delta_{max}}$. We find $R_{tot} = \frac{1}{4}\sigma(\omega\mu)^2\pi N_a^2\hat{\delta}^3 = \frac{\pi}{2}\omega\mu N_a^2\frac{\hat{\delta}^3}{\delta^2} = \alpha\frac{\hat{\delta}^3}{\delta^2}$. (This analysis is only intended to give a qualitative understanding and should not be taken too seriously.) The average absorbed power is

$$\bar{P}_{\text{tot}} = \frac{1}{2} I_s^2 R_{\text{tot}}. \tag{3.41}$$

The power is proportional to the effective resistance of the plasma – the external circuit does not seem to play a major role in the feedback. The modification to the skin depth was needed to make the resistance have a maximum. This resistance is proportional to

$$\hat{\delta}^3/\delta^2 = \frac{\delta \delta_{\text{max}}^3}{(\delta + \delta_{\text{max}})^3}. \tag{3.42}$$

The maximum of this quantity comes when

$$\frac{1}{(\delta + \delta_{\text{max}})^3} - \frac{3\delta}{(\delta + \delta_{\text{max}})^4} = 0 \tag{3.43}$$

or

$$3\delta = \delta + \delta_{\text{max}}; \text{ so } \delta = \delta_{\text{max}}/2. \tag{3.44}$$

This means that for maximum power deposition the skin depth should be half of the vacuum skin depth.

This discussion illustrated several points in addition to estimating how much power is deposited. One was that we made a series of estimates of which effects mattered and reached at least one dead end before a reasonable physical picture emerged.

The electrical characteristics of a slightly different inductive discharge are investigated in Ref. [60].

3.4 Circuit Model of a Capacitive Discharge

In an rf discharge, which is driven by applying a voltage to a pair of electrodes, the behavior of the sheaths next to the electrodes is particularly important. When a time-varying voltage is applied to an electrode the sheath thickness alters, primarily because the electrons move toward or away from the electrode. If the variation in voltage is oscillating at a radio frequency, the ions will not be able to respond fast enough to change their density much during an rf cycle, so nearly all the response is due to the electrons. When the voltage on an electrode gets more negative the electrons in the plasma are repelled further from the electrode, thereby exposing more of the positive charge at the edge of the sheath next to the plasma.

Because there are very few electrons in the sheath and because the ions move slowly little conduction current usually flows in the sheath. The extra ions that are exposed if the electrons move back from the electrode are a source of electric field lines that go to the electrode. There must be enough negative charge on the electrode for the field lines to end on. The field lines can (for the most part) not go into the plasma because it is a good conductor that supports only very small electric fields. The electrons are free to move around so as to shield the interior of the plasma from electric fields. Because electric field lines are not likely to go into the plasma, exposing the extra positive charge at the plasma/sheath edge causes more field lines to go through the sheath to the electrode. The extra electric field lines in the sheath increase the sheath electric field, the sheath thickness, and the voltage in the sheath. The changing electric field gives a displacement current density

$$\mathbf{j}_D \equiv \frac{\partial \mathbf{D}}{\partial t}. \tag{3.45}$$

The fact that the current in the sheath is largely displacement current implies that the sheath behaves in some ways like a variable-thickness capacitor.

The total current consisting of displacement plus conduction current must be the same in the plasma as in the sheath. When we move from sheath to "bulk" plasma the current switches from being displacement current to conduction current. The bulk plasma must therefore tend to behave like a resistor, although the behavior of the resistor is nonlinear.

It may be useful, if they are available, to use tables of plasma properties such as current density and ionization rate versus E/N to understand the behavior of the "resistor." The energy an electron with charge e picks up from the electric field E in traveling a mean free path λ in the direction the field pushes the electron is $\Delta \epsilon = e E \lambda$, and since $\lambda = \frac{1}{N\sigma}$ the energy $\Delta \epsilon$ is proportional to E/N. Because many plasma parameters depend on the mean energy and because this depends on $\Delta \epsilon$ the plasma parameters are frequently measured and tabulated as functions of E/N.

The circuit model that we have (Fig. 3.2) consists of a resistive main body of the plasma with sheaths, which resemble capacitors, between the plasma and the electrodes. There is one situation where this picture breaks down. The sheaths are like capacitors because the electrons are pushed out of the sheaths by the negative charge on the wall. However, if the external circuit removes the negative charge on the electrode, the sheath collapses and electrons can reach the wall in large numbers [61]. Usually the electrode is negative relative to the plasma because the potential is set up to keep electrons off the electrode. When the sheath collapses the sheath voltage is zero. The fact that the capacitor collapses and becomes a short circuit whenever the voltage across the sheath fails to hold out electrons can be modeled with a diode in parallel with the capacitor. The resistor R_s allows for the ion current through the sheath.

The direction of the diodes is chosen to allow electrons to go out to the electrode when the electrode is not negative relative to the plasma. The diodes also make sure that the electrode is not positive relative to the plasma, since in reality this would rapidly pull the electrons out of the plasma and make the plasma more positive again.

The electron current to the electrode is very large if the sheath fails to repel the electrons from the electrode. In many cases the sheath can only collapse very briefly because such a large electron current will tend to make the plasma positive relative to the electrode.

The most common arrangement of the electrodes has a relatively small powered electrode, with the grounded metal wall of the vacuum chamber acting as the other. A large current will flow to one or the other electrode, unless (i) the plasma

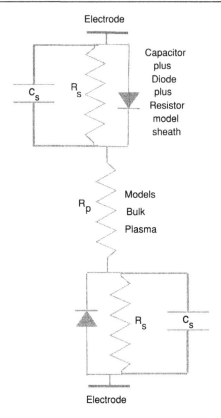

Fig. 3.2. Equivalent circuit for a capacitive discharge.

potential is somewhat higher than the potential of either electrode all the time or (ii) the potential only very briefly becomes about as low as the electrode potential. If the chamber is grounded and the powered electrode voltage is $\Phi_{pe} = \Phi_o \cos \omega t$, then the plasma can float at a potential $\Phi_p \geq \Phi_o$. Alternatively, if there is a capacitor in the external circuit a dc voltage can be supported between the electrodes. The mean voltage of the powered electrode often has a dc offset of $-\Phi_o$; hence $\Phi_{pe} = \Phi_o(\cos \omega t - 1)$. This allows the plasma to float at a potential just above the grounded wall potential without ever dropping below the powered electrode potential. The very large area of the wall relative to the powered electrode makes it likely that the plasma will be at a potential close to the wall potential; consequently, this result is often expected.

For experiments on impedance components of an rf discharge, see Ref. [62].

3.5 Circuit Model of an Electron Cyclotron Resonance Discharge

An Electron Cyclotron Resonance (ECR) discharge (Fig. 3.3), is heated by micro-waves [63]. The electrons gyrate around magnetic field lines, and, at one surface within the plasma, their gyration frequency equals the microwave frequency, enabling the electrons to be resonantly heated by the microwaves.

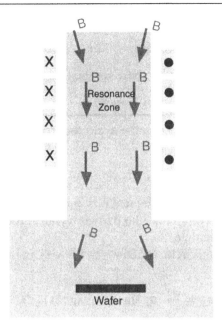

Fig. 3.3. Schematic of an ECR discharge.

Charged Particle Motion in a Magnetic Field

To understand the ECR discharge we need to know how the magnetic field affects electrons and ions. The force exerted by the electric field $\mathbf{E}$ and magnetic field $\mathbf{B}$ on a particle with charge q moving with velocity $\mathbf{v}$ is the Lorentz force:

$$\mathbf{F} = q(\mathbf{E} + \mathbf{v} \times \mathbf{B}). \tag{3.46}$$

We can find the magnetic force parallel to $\mathbf{B}$ and perpendicular to it. We consider a straight magnetic field. The parallel magnetic force is zero and so if $E_\parallel = 0$,

$$m\frac{dv_\parallel}{dt} = 0, \tag{3.47}$$

but the perpendicular magnetic force is $q\mathbf{v} \times \mathbf{B} = q\mathbf{v}_\perp \times \mathbf{B}$; so

$$m\frac{d\mathbf{v}_\perp}{dt} = q\mathbf{v}_\perp \times \mathbf{B}, \tag{3.48}$$

where $(v_\parallel, v_\perp)$ are the parallel and perpendicular components of the velocity. The straight magnetic field allows particles to move along its direction without $v_\parallel$ being affected. The perpendicular velocity $v_\perp$ is made to gyrate. Each component of $v_\perp$ oscillates but they are out of phase with each other by $\pi/2$. If we assume that $v_\perp$ can be written as

$$\mathbf{v}_\perp = \mathbf{a}_x v_{xo} \cos \omega_c t + \mathbf{a}_y v_{yo} \sin \omega_c t \tag{3.49}$$

and $\mathbf{B} = B_o \mathbf{a}_z$, the result for the x component of the force is

$$-m v_{xo} \omega_c \sin \omega_c t = q v_{yo} B_o \sin \omega_c t \tag{3.50}$$

and for the y component

$$mv_{yo}\omega_c \cos \omega_c t = -qv_{xo}B_o \cos \omega_c t. \tag{3.51}$$

For a positive particle we can solve these for the frequency ω_c to get

$$\omega_c = qB_o/m, \tag{3.52}$$

which is called the gyrofrequency or the cyclotron frequency. We also have $v_{yo} = -v_{xo}$; so if we set $v_{xo} = v_\perp$ we get

$$\mathbf{v}_\perp = v_\perp(\mathbf{a}_x \cos \omega_c t - \mathbf{a}_y \sin \omega_c t). \tag{3.53}$$

For a negative particle, the gyrofrequency is equal to the magnitude of qB_o/m and the velocities obey $v_{yo} = v_{xo}$; the particle gyrates in the opposite direction, compared to a positive particle.

The particle traces out a circle. Since $\mathbf{v}_\perp \equiv \frac{d\mathbf{r}_\perp}{dt}$ we can integrate with respect to time and find

$$\mathbf{r}_\perp = \frac{v_\perp}{\omega_c}(\mathbf{a}_x \sin \omega_c t + \mathbf{a}_y \cos \omega_c t). \tag{3.54}$$

The magnitude of this vector is the gyroradius $r_\perp \equiv v_\perp/\omega_c$. This is the radius of the circle traced by the particle's orbit. For the magnetic fields used in the ECR discharge the electron gyroradius usually is a fraction of a millimeter. Ions have a much larger $r_\perp$ because of their much greater mass. Because of the small electron gyroradius we speak of electrons being "tied" to the field line.

When particles have collisions their gyration is interrupted, and when they start to gyrate again they are moving in a different direction so they orbit around a different field line. The distance between the old and new field lines is typically about as big as the gyroradius. Hence collisions can lead to a random walk between field lines with a step size of $r_\perp$. In most plasmas, without a magnetic field, electrons escape much faster than ions. However, in the diffusion across magnetic field lines the small $r_\perp$ for electrons means that ions cross field lines in less time than electrons. In an ECR discharge, electrons can leave rapidly along field lines even if this means traveling a long way. Because ions get out fairly rapidly across the field lines, they have the option of going the shorter distance to the radial wall or all the way to the end of the discharge, along the field lines.

Circuit Model of the Plasma and the Surface

From a circuit point of view, the electrons in the center of the plasma see a small impedance along the field lines to the end of the discharge and a large impedance across the field lines to the walls. The ions probably see a slightly smaller impedance between the center and the walls than between the center and the ends, because of the shorter distance to the walls, which compensates for the magnetic field impeding the cross-field motion. If the walls are conducting the electrons will leave to the ends, ions will leave to the side walls, and a current will flow along the wall to allow the electrons to neutralize the ions. If the walls are insulating this cannot happen and electric fields would be expected to be set up to equalize the fluxes of electrons and ions to each point on the wall. The boundary condition at the wall is critical to how the electric fields will respond. In practice the insulation is often a thin layer of insulator over a metal surface and there may be some questions about how good the insulation really is.

3.6 Circuit Model for Damage Studies

Damage may be caused by energetic ions striking the substrate. In high-density reactors such as the ICP [64, 65] and ECR [63] discharges, where the plasma number density is usually well above 10^{10} cm^{-3}, and may be 10^{12} cm^{-3} or more, the processing usually does not need very high sheath voltages and so the ion energies are less and damage mechanisms other than impact by energetic ions are more important. Of particular concern when fabricating or patterning insulating layers on the surface of the wafer is the build up of charge on the layer, leading to a high voltage across the layer, which could cause electrical breakdown.

To see when a high voltage is liable to be set up across a surface layer (such as a gate dielectric) we can consider an equivalent circuit where the insulating structures on top of the wafer are treated as individual capacitors. It must be kept in mind that these circuit models are only extended analogies and cannot be taken very seriously, however. Fortunately, only the capacitors are essential in our analysis.

The silicon itself typically has a high enough conductivity to be considered as a good conductor, connecting the back sides of all the capacitors on the silicon surface to the external circuit. The external circuit in turn has a blocking capacitor in series with the connection to the back of the wafer, as well as an rf voltage in series before the connection to the conducting wall that surrounds most of the plasma. The final series circuit element is the plasma itself before we again reach the wafer. Although the sheaths behave as capacitors with respect to rf voltages, in dc conditions the sheaths do not have a displacement current in them and they can always carry a small conduction current; thus we must be careful about treating sheaths simply as capacitors.

To understand one reason why damage might occur due to charging we have to consider how the plasma interacts with the surface. When the plasma first reaches the surface the electrons begin to charge the surface negatively. The potential Φ_s that is set up across the sheath in steady state repels nearly all of the electrons. To do this the mean potential energy $e\Phi_s$ required for an electron to reach the surface must be several times as big as the mean electron energy. In addition, the dc sheath voltage must be high enough to prevent rf voltages from making the wafer surface attractive to electrons at any point in the cycle.

The circuit shown in Figure 3.4 represents some of the features of the plasma, the wafer, and the vacuum chamber that we have described. The plasma is represented by a network of resistors. The critical aspect of the model is the sheath adjacent to the wafer. In addition to the capacitor and diode that were in the previous model of the rf discharge we have added a resistor in series with a dc voltage source. This resistor plus voltage source allow a current to flow until the surface charges up.

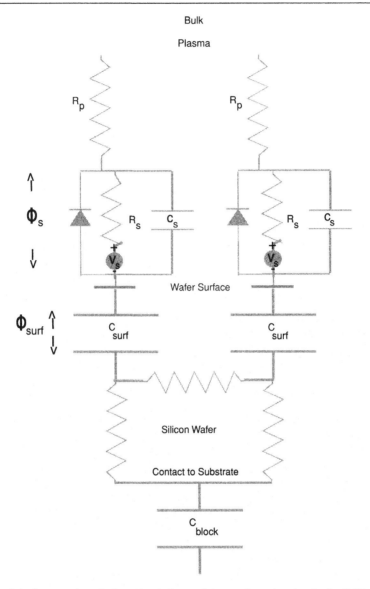

Fig. 3.4. Suggested equivalent circuit for studying surface charging in the ECR discharge.

After the surface has charged to a steady-state level, the electron and ion currents to the surface are equalized on average over a cycle. The dc voltage source in series with the resistor is necessary in part to model the fact that electron current flows to the surface even when the sheath potential pushes electrons away. When high-energy electrons can still get over the energy barrier faster than ions are pulled to the sheath there is a net flux of negative charge to the surface. Making the potential barrier higher can equalize the fluxes of ions and electrons and making it greater still can make the ion flux greater than the electron flux.

In the circuit model the voltage applied to the sheath is applied to the series combination of resistor and voltage source. This applied voltage Φ_s provides the "energy barrier" $e\Phi_s$ that the electrons must overcome to get to the wall. When Φ_s is large the electrons are repelled and ion current dominates. In the circuit, when Φ_s is larger than the voltage of the voltage source a positive current flows through the resistor to the surface. Similarly, when Φ_s is small the circuit gives a current away from the surface. Finally, we must also keep in mind that most of the electrons reach the surface when the rf voltage source is most positive which is when the voltage source raises the potential at the surface to its highest value. This is discussed more, later in the section.

The diode makes sure that when the rf voltage reaches its maximum value, the sum of the voltages between ground and the sheath edge immediately above the wafer cannot exceed the potential of the plasma opposite that point on the wafer. In other words the diode makes sure Φ_s is not negative.

If we assume that the sheath voltage Φ_s is only zero instantaneously we can analyze the circuit while largely ignoring the diode. Suppose the impedance of the blocking capacitor is small compared to that of the insulating structures and of the sheath so that we can ignore the blocking capacitor also. It may be useful to examine the section of circuit containing only one capacitive structure on the surface.

Φ_{plasma} is the potential the plasma settles down to in the presence of the grounded tank. The wafer is probably too small in area to affect Φ_{plasma} much and so Φ_{plasma} in volts is usually a few times the electron temperature T_e in electron-volts. If the barrier between plasma and wall is $e\Phi_{\text{plasma}} \sim 3k_B T_e$, most electrons are repelled by the wall. This potential is required in steady state to keep the electron current to the wall to a moderate value.

If Φ^{rf} has a large amplitude then $|\Phi^{\text{rf}}| \gg |\Phi_{\text{plasma}}|$ and we can set Φ_{plasma} to zero.

We can use superposition to split the circuit into rf and dc components (Figs. 3.5 and 3.6), provided we make sure that the total Φ_s is never negative, since the only nonlinear element in the circuit is the diode, which would turn on to prevent Φ_s from going negative.

The dc circuit implies that the voltage source in the sheath has a voltage equal in magnitude to the dc voltage across the surface structure:

$$\Phi_{\text{surf}}^{\text{dc}} = \Phi_s^{\text{dc}}. \tag{3.55}$$

The impedance of the resistor in the rf circuit is probably very high. Thus it can be ignored and the voltage drop across the sheath can be obtained by capacitive voltage division. The rf voltage is dropped across the sheath and the surface structures and is divided in proportion with the impedance. Therefore

$$\Phi_s^{\text{rf}} = \Phi^{\text{rf}} \frac{(\omega\, c_s)^{-1}}{(\omega\, c_{\text{surf}})^{-1} + (\omega\, c_s)^{-1}} = \Phi^{\text{rf}} \frac{c_{\text{surf}}}{c_s + c_{\text{surf}}} \tag{3.56}$$

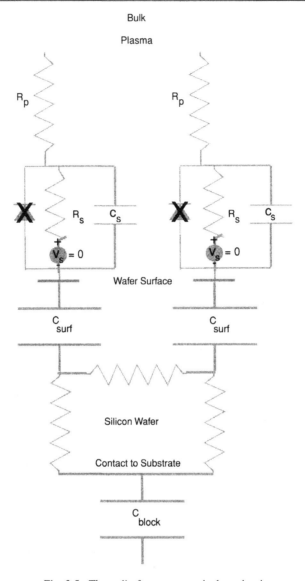

Fig. 3.5. The radio frequency equivalent circuit.

and

$$\Phi_{\text{surf}}^{\text{rf}} = \Phi_{\text{rf}} \frac{c_s}{c_s + c_{\text{surf}}}. \tag{3.57}$$

To keep Φ_s from going negative we need the dc sheath voltage plus the rf sheath voltage to have a maximum value of zero.

$$\Phi_s^{\text{rf}} - \Phi_s^{\text{dc}} = 0. \tag{3.58}$$

This and the result from the dc circuit mean

$$\Phi_{\text{surf}}^{\text{dc}} = \Phi_s^{\text{dc}} = \Phi_s^{\text{rf}}. \tag{3.59}$$

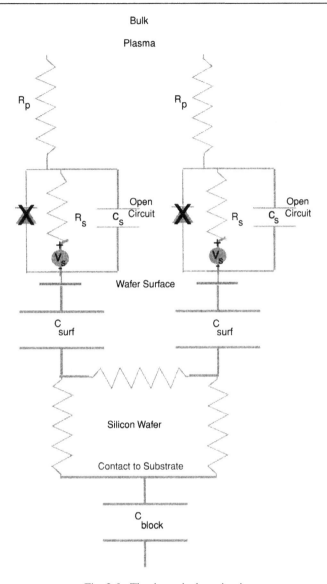

Fig. 3.6. The dc equivalent circuit.

The voltage across the surface is

$$\Phi_{\mathrm{surf}} = \Phi_{\mathrm{surf}}^{\mathrm{dc}} + \Phi_{\mathrm{surf}}^{\mathrm{rf}} \cos \omega t. \qquad (3.60)$$
$$= \Phi_{\mathrm{s}}^{\mathrm{rf}} + \Phi_{\mathrm{surf}}^{\mathrm{rf}} \cos \omega t.$$

The maximum voltage across the surface structure is $\Phi_{\mathrm{s}}^{\mathrm{rf}} + \Phi_{\mathrm{surf}}^{\mathrm{rf}} = \Phi^{\mathrm{rf}}$; consequently, at some instant in the cycle the entire rf voltage appears across the surface structure. Because the diode limits the voltage across the sheath to be zero when the rf power supply is most positive, the surface must be blocking the rf voltage at this instant. This confirms that the entire amplitude of Φ^{rf} appears across the surface structure at this point in the cycle.

Exercises

1. Criticize the analysis of the ICP that led to the expressions for the impedance seen in the primary circuit. What is wrong with assuming a current sheet? Where does the voltage in the primary actually appear? (Think about an antenna that is a very good conductor.)

2. Rederive the expressions for the real and imaginary parts of the impedance seen in the primary circuit of an ICP, allowing for a complex conductivity of the plasma.

3. The "maximum power theorem" indicates the impedance values for which the maximum power is transmitted from a power supply to a load. The power supply used for an inductively coupled plasma is usually modeled as a current source, in parallel with a large impedance. If the plasma behaves like an inductor in series with a resistor, what "matching network" will provide maximum power transfer?

4. Look up the electrical and thermal conductivities of silicon – see, for example Ref. [29]. Note that the electrical conductivity may have to be estimated from a mobility. Consider intrinsic silicon and silicon doped to be n-type with a dopant density of $N_D = 10^{19}$ cm^{-3}. Assume a temperature of $T = 300$ K.

   For a wafer of thickness $t_w = 1$ cm, width $W_w = 5$ cm, and length $L_w = 20$ cm, what is the resistance R_w between the ends, which are 20 cm apart? What combination of t_w, W_w, and L_w does R_w depend on? Explain why R_w is usually quoted as ohms per square. How many squares is this resistor made up from? (Consider both intrinsic and doped silicon.)

   Suppose this wafer is in contact with a plasma that sustains a potential difference of $0.1\,T_e$ volts between its ends, where T_e is the electron temperature in electron-volts; $T_e \simeq 5$ eV. What current flows in the wafer? How much power is dissipated in the wafer? (Consider both intrinsic and doped silicon.)

   If a temperature difference of $10^{-3}\,T_e$ is sustained between the ends of the wafer, meaning that since one electron volt is roughly 10^4 K, $10^{-3}\,T_e$ is 50 K, how much power flows from one end of the wafer to the other?

5. Criticize the circuit model introduced to study damage to the surface being processed. Estimate the values of the components and check to see if they are as large/small as was claimed. Sketch the voltage across the sheath as a function of time. Repeat for the voltage across the surface structure. Are the conclusions of the analysis consistent with your sketch? Is your sketch physically reasonable?

4

Plasma Models

4.1 Plasma Models

In this chapter we begin to develop more accurate descriptions of the same plasma reactors we considered before, which go beyond the circuit models used so far. The information we can extract from circuit models is useful in some contexts but is very limited. We begin with a discussion of a parallel-plate rf discharge, which is normally operated at higher neutral pressures than the other reactors we consider. The behavior of the plasma in inductively coupled plasmas will be outlined next. Ambipolar diffusion is introduced, to help in the description of the ICP. The ECR discharge is made more difficult to describe by the presence of the magnetic field, which steers electrons almost exclusively along the magnetic field lines but which impedes ions only moderately on their way to the wall. ECR plasmas are considered next. Calculations of the electron and ion distribution functions are the last major topic in this chapter. Before discussing the rf discharge we will describe a type of experiment that can provide a very useful basis for comparison with more complicated plasmas. These experiments involve "swarms" of electrons subjected to an electric field that is constant in space and time.

4.2 Swarm Measurements

Swarm experiments [66] involve a so-called swarm of electrons with a low enough density so that their charge does not (significantly) alter the electric field. The conditions for this to occur were discussed in the section on plasma electrostatics. Because the electric field in this sort of experiment can be kept constant in space and time, the behavior of the electrons can be studied for a clearly specified field. The response of the electrons in this situation depends on the value of E/N. We shall now describe why this might be expected.

The energy a particle acquires in going a distant equal to the mean free path λ is proportional to the mean free path λ multiplied by the electric field E. After going a distance λ and suffering a collision, the particle will normally have its direction randomized (at least partially). Similarly, the inelastic mean free path λ^{inel} multiplied by E is proportional to the amount of energy the particle would pick up before it would be likely to lose energy in an inelastic collision, if the particle moved in a straight line between inelastic collisions. The typical amount of energy the electrons acquire will in any case be proportional to the mean free path (*mfp*) multiplied by E. Any mean free path is (usually) inversely proportional to the density of background particles that are being collided with. In the plasmas we are studying the electron *mfp* usually varies as $1/N$, where N is the neutral density. This means that any *mfp* multiplied by E is proportional to E/N; so the amount of energy electrons pick up before any sort of collision varies as E/N. The mean energy of the electrons is consequently expected to depend on E/N instead of just E [67]. Many other properties of the electrons also depend on E/N, such as the fraction of the electrons that are hot enough to ionize neutral atoms. The mean energy, the ionization rate per electron, and the mean velocity in the direction of the electric field are all known and available for various neutral gases and are tabulated in terms of the value of E/N [68].

Consider a discharge where the electron behavior is dominated by the electrons being heated by a roughly uniform electric field. It will be possible to predict the electron properties as functions of E/N, if swarm data are available for the particular neutral gas. As we shall see in the next section, it happens frequently that a discharge will undergo a transition as we vary an external parameter such as the power supplied. For one range of parameters the electrons may be mostly heated locally by a locally near-uniform E field inside the plasma so swarm data are applicable. For another range of parameters the heating might happen at the edge of the plasma, producing hot electrons, which move into the plasma interior and cause ionization. The situation where electrons are being heated in one place and are moving a long way before they ionize neutrals is referred to as "nonlocal" heating. The difference between local and nonlocal heating can lead to important variations in the density profiles of charged particles and in the electron mean energy and distribution function. The parallel plate discharge provides a good example of a transition from local to nonlocal heating, which occurs as the applied voltage is varied.

4.3 Capacitive rf Discharges

In Chapter 3 on circuit models of discharges we described the rf capacitive discharge as a series combination of a capacitor, a resistor, and another capacitor. The capacitors represent the sheaths, which are regions of strong electric fields between the walls and the plasma. The resistor represents the main part of the plasma, which is nearly neutral. The sheaths resemble capacitors because there is usually little conduction current in the sheath but a substantial displacement current. The main plasma volume has weak electric fields, and hence little displacement current, but large conduction currents, and so it is more like a resistor. Plasma heating can take place in this resistor, if the electric field is not too small. This can be a "local" heating mechanism, since the local electric field controls the ionization rate in the main plasma. The sheaths physically expand and contract during the rf cycle and the strong fields in and near the sheath provide a second means to create a population of hot electrons, if the expanding sheath catches up with enough electrons. When electrons heated in the sheath subsequently move into the main plasma and ionize neutrals, they provide nonlocal heating in which the sheath electric field produces ionization in the bulk plasma.

We have identified two possible heating mechanisms so far in capacitive rf discharges. Local heating in the plasma can be expected to produce significant ionization if E/N is large enough. Nonlocal heating could occur if the sheaths can heat enough electrons near the sheath edge, which would then move back into the main plasma and cause ionization. It is interesting to investigate which of these mechanisms is most important and whether there can be a changeover in which dominates. This transition should take place at some value of the externally applied voltage, because experiments show a dramatic change in behavior as the voltage is varied.

4.3.1 Local Heating in Capacitive Discharges

To estimate the heating rate by each mechanism, we begin by examining the displacement current in the sheath, which is very nearly the total current in the sheath. This in turn must be equal to the current in the plasma, which is very nearly all conduction current. The conduction current will in turn allow us to find the local heating rate. To find the displacement current, we need to know the electric field in the sheath. This field is determined by the applied voltage and the charge density in the sheath. The ion density in the sheath n_s is typically an order of magnitude less than the peak ion density in the center of the plasma n_m, that is, $n_s = f_s n_m$, where $f_s \sim 0.1$. (This difference in density can be explained in terms of the ion speeds and the ion fluxes in the two locations; for now we will claim that it is a typical experimental situation.) There are almost no electrons in the sheath, so (if n_s is roughly constant) Poisson's equation reads

$$\frac{d^2\Phi}{dz^2} = -\frac{\rho}{\epsilon_o} = -\frac{n_s e}{\epsilon_o} = -\frac{f_s n_m e}{\epsilon_o}. \tag{4.1}$$

This means that in the sheath

$$E = -\frac{d\Phi}{dz} \simeq \frac{f_s n_m e}{\epsilon_o}(z - z_s). \tag{4.2}$$

Here z_s is the location of the sheath–plasma boundary, where E goes to zero, and $z = 0$ at the left electrode.

At the left electrode and in the sheath by this electrode $z < z_s$ and E is negative. At the right electrode z_s has a different value, which is larger than the z_s for the left electrode. In the right sheath $z > z_s$ and so E is positive. These electric fields have the correct sign to push ions out and to hold electrons in. To find z_s we have to calculate Φ in terms of z_s. Integrating again we have the potential in the sheath, measured relative to the main plasma, which is taken to be at zero potential:

$$\Phi = -\frac{f_s n_m e}{2\epsilon_o}(z - z_s)^2. \tag{4.3}$$

The voltage Φ has a sinusoidal component, but it also has a dc component, which is needed to keep the plasma from going negative relative to the electrode.

The voltage at the electrode is $\Phi(z = 0)$ and this voltage must be less than or equal to zero:

$$\Phi(z = 0) = -\Phi_{dc} - \Phi_{rf}\cos\omega t = -\frac{\alpha}{2}z_s^2, \tag{4.4}$$

where

$$\alpha = \frac{f_s n_m e}{\epsilon_o}. \tag{4.5}$$

If we keep $\Phi(z = 0) \leq 0$ by choosing $\Phi_{dc} \simeq \Phi_{rf}$ then

$$-\Phi(z = 0) = \Phi_{rf}(1 + \cos\omega t) = \frac{\alpha}{2}z_s^2 \tag{4.6}$$

and so

$$z_s = \left[\frac{2\Phi_{rf}}{\alpha}(1 + \cos\omega t)\right]^{1/2}. \tag{4.7}$$

The electric field at $z = 0$ is $E(z = 0) = -\frac{d\Phi}{dz} = -\alpha z_s$. Consequently, the displacement current density $\frac{\partial D}{\partial t} = \epsilon_o \frac{\partial E}{\partial t}$ is

$$\frac{\partial D}{\partial t} = -\epsilon_o \alpha \frac{\partial z_s}{\partial t} = \epsilon_o(2\alpha\Phi_{rf})^{1/2}\frac{\omega\sin\omega t}{2(1 + \cos\omega t)^{1/2}}. \tag{4.8}$$

When z_s is close to zero we need to examine this expression carefully, to see what happens. If $\omega t = \pi + \omega\delta t$, then $\sin(\pi + \omega\delta t) \simeq -\omega\delta t$ and $\cos(\pi + \omega\delta t) \simeq -(1 - \frac{1}{2}(\omega\delta t)^2)$. Then $\sin\omega t / (1 + \cos\omega t)^{1/2} \simeq \frac{\omega\delta t}{\sqrt{\frac{1}{2}}\omega\delta t} = \sqrt{2}$.

From this we conclude that the displacement current density is of the order of

$$J_D^o = \omega\epsilon_o(\alpha\Phi_{rf}/2)^{1/2}. \tag{4.9}$$

The conduction current in the bulk must be equal to the displacement current at the electrode edge of the sheath, if there is no displacement current in the bulk and no conduction current at the electrode surface. To find the conduction current we need to know the drift velocity v_e^{dr} of electrons in the applied field. The mobility of electrons is $-\mu_e$ and is defined so that

$$v_e^{dr} = -\mu_e E. \tag{4.10}$$

Hence the conduction current density is

$$J = -n_e e v_e^{dr} = n_e e \mu_e E^b, \tag{4.11}$$

where E^b is the field in the bulk of the plasma.

Since this has to be equal to the sheath displacement-current density, we have

$$n_e e \mu_e E^b \simeq \omega\epsilon_o(\alpha\Phi_{rf}/2)^{1/2}. \tag{4.12}$$

From this we can predict how E/N might be expected to vary. The mobility $-\mu_e$ is inversely proportional to N (the mobility goes down as the neutral density goes up); so $\mu_e E$ is proportional to E/N. From the above, we have

$$\mu_e E^{\mathrm{b}} \simeq \frac{\omega\epsilon_o}{n_e e}(f_s n_m e\Phi_{\mathrm{rf}}/2\epsilon_o)^{1/2}. \qquad (4.13)$$

In the bulk, the electron density n_e is equal to the ion density. Then the maximum density $n_{\mathrm{m}} = n_e$ at the peak of the density and

$$\frac{E^{\mathrm{b}}}{N} \sim \mu_e E^{\mathrm{b}} \simeq \omega\sqrt{\frac{\epsilon_o f_s \Phi_{\mathrm{rf}}}{2n_m e}}. \qquad (4.14)$$

The first conclusion from this is that if n_{m} goes up with applied voltage Φ_{rf} faster than Φ_{rf} itself increases (which it does) then E^{b} goes down as the plasma density increases. In part this is because the E^{b}/N required to drive a given current goes down as the electron density goes up. If E^{b}/N goes down with increasing plasma density then local heating will also be less effective at high plasma density. If local heating is ineffective then, if the discharge is sustained at all, it must be by nonlocal heating. We now have a prediction that we might expect a transition as Φ_{rf} increases, provided n_{m} increases faster than Φ_{rf}, from local heating at low Φ_{rf} and low plasma density to nonlocal heating at high Φ_{rf} and high plasma density.

4.3.2 Nonlocal Heating in Capacitive Discharges

The argument given in the last section only establishes that local heating is not effective if $\Phi_{\mathrm{rf}}/n_{\mathrm{m}}$ is small. The next question might be, when is nonlocal heating effective? To estimate a heating rate due to the moving sheaths we need to know how many electrons are affected by the sheaths and how much energy they might pick up from the sheath. The maximum number of electrons in the region affected by the moving sheath (per unit area) is roughly the density of charged particles in the sheath multiplied by the maximum sheath thickness:

$$N_{\mathrm{s}} \simeq n_{\mathrm{s}} z_{\mathrm{s}} \simeq f_{\mathrm{s}} n_{\mathrm{m}} z_{\mathrm{s}}. \qquad (4.15)$$

If these particles simply bounced off the front of the moving sheath as if it were a rigid wall, the kinetic energy they acquired would be of the order of $T = \frac{1}{2}mv_{\mathrm{s}}^2$ where v_{s} is the speed of the sheath edge. Since v_{s} is of the order of ωz_{s}, $T \sim \frac{1}{2}m\omega^2 z_{\mathrm{s}}^2$. The sheath pushes out with frequency $\omega/2\pi$ and we shall suppose that each time it does so it gives an energy T to a number of particles equal to the number in the sheath, N_{s}, which is the density in the sheath n_{s} multiplied by the sheath thickness z_{s}. The heating rate per unit area of the sheath (as viewed from the electrode next to it) is thus

$$\omega_{\mathrm{nonlocal}} \sim \frac{\omega}{2\pi}(f_s n_m z_s)\left(\frac{1}{2}m\omega^2 z_s^2\right) \sim f_s \frac{n_m m}{4\pi}(\omega z_s)^3. \qquad (4.16)$$

Using $z_s \sim (2\epsilon_o \Phi_{rf} / f_s n_m e)^{1/2}$ gives

$$\omega_{nonlocal} \sim f_s \frac{n_m m}{4\pi} \omega^3 \left(\frac{2\epsilon_o \Phi_{rf}}{f_s n_m e} \right)^{3/2}. \tag{4.17}$$

Expressed as a heating rate per electron, by dividing this heating rate by the total number of electrons per unit area of the sheath/electrode, which is roughly $n_m L$, this rate also depends on Φ_{rf}/n_m. (L is the length of the discharge and all the "total" numbers we calculate are per unit area of the electrode.) However, when we examined the local heating we found that E^b/N varied as $(\Phi_{rf}/n_m)^{1/2}$. The local heating rate per electron and the associated ionization rate, according to swarm data vary more or less exponentially with $E^b/N \sim (\Phi_{rf}/n_m)^{1/2}$, which is a more rapid variation. This implies that at low Φ_{rf}/n_m the heating will be nonlocal, but local heating can indeed be expected to dominate at high Φ_{rf}/n_m.

To show that the sheath mechanism for nonlocal heating, or the local bulk heating, is likely to be able to sustain the discharge under any circumstances, we would need to do much more extensive calculations than these [69, 70]. Numerical calculations do, in fact, confirm this picture of a transition between types of heating.

If Φ_{rf}/n_m (and hence E^b/N) is large all the electrons in the bulk are heated directly by the electric field and so they are likely to have a high average energy. If, however, the sheath heating is the only way to keep the discharge going because E^b/N is small, most of the bulk electrons will have a low average energy. The "hot" electrons produced by the moving sheath can provide all the ionization that is needed if the nonlocal heating is effective. In high plasma density discharges such as the ICP, Coulomb collisions, which are collisions between charged particles, transfer kinetic energy from the hottest electrons to the cold group of electrons and keep the distribution close to a Maxwellian. In the capacitive discharge the plasma density is usually too low to allow the Coulomb collisions to transfer the energy effectively. The cold electrons form the vast majority of the electron population and their mean energy will be quite low. The importance of this is that, although hot electrons provide the ionization, it is the entire population of electrons that control the electric fields by shielding the bulk plasma, moving around in response to applied electric fields. The potentials set up in the plasma are of the order of the electron temperature T_e (in eV) and if most of the electrons are very cold the potential $\Phi \sim T_e$ will be very small. In turn, if Φ is small the electric fields that pull ions out of the plasma are weak and thus ions take a long time to leave; the ion confinement time is high and so the ion density is high. So even if the ionization rate is not very high the absence of local heating causes high densities because it leads to a long ion confinement time. This effect of creating a population of "cool" electrons helps to explain why the nonlocal heating can occur at high electron densities, whereas the local heating is a low electron density phenomenon.

This discussion is typical of analytic plasma models in that it is plausible but not entirely convincing. When multiple analytic estimates are made, each relying on the one before, errors can grow extremely rapidly. This means that we have to draw frequently on experimental observations to stay on track.

4.4 Analytic Model of Transport in an ICP

This section examines two aspects of the power input and particle production in ICPs: the location of the heating and the transport around the rest of the discharge of the electrons that are heated.

Role of the Skin Depth

Inductively coupled plasmas (ICPs) used for semiconductor processing tend to have neutral pressures in the range from about 1 mTorr to about 100 mTorr. The electron density in these discharges tends to be high, making the conductivity σ also high. The inductive coupling of the plasma to the current in the antenna means that an image current flows in the plasma. The magnitude of the current would equal the primary current in the ideal case; in reality it is proportional to the primary current. The current density J equals σE and this current density flows in a skin depth [71] $\delta \sim \sigma^{-1/2}$. The current $I \sim \sigma E \delta \sim \sigma^{1/2} E$; so if the current is constant an increase in σ causes a decrease in E. At the lowest neutral pressures E/N is small inside the plasma despite the small value of the neutral density N because N being small makes σ large. The electron density n is typically large, making σ larger still. The large conductivity σ means that E is very small. However, at high neutral pressures E/N can be large despite N being large because the conductivity is low and E can be large. Thus we might have a local heating regime at high neutral pressure but must look for nonlocal heating (where the electrons have to get very close to the antenna, where n and σ are small, to be heated) at low neutral pressure.

In the argument just given we considered E/N within a distance from the antenna equal to the skin depth δ. Outside δ there is very little field in an ICP, so even if E/N is large enough to cause local heating in the skin depth, it will not be able to do so elsewhere.

Random Walk of Electrons

Another issue critical for determining the nature of the heating is that of the relative size of the electron mean free path compared to the system dimension L. At 1 mTorr the electron elastic *mfp* is a little less than the system size of a typical ICP reactor. At slightly higher pressures the electrons suffer multiple elastic collisions in crossing the discharge. Thus their direction is randomized but they are not likely to lose much energy. If they can cross the entire discharge before they suffer an inelastic collision, the ionization when it does take place can be anywhere in the volume.

When the electron has multiple elastic collisions in crossing the discharge it diffuses across the distance L. The net distance away from the starting point that is traveled in N steps of a random walk of step size λ (in this case λ_e^{el}, the elastic mean free path) is about $\sqrt{N}\lambda$. To travel outward through the distance L the electron must take N steps, where $L \simeq \sqrt{N}\lambda$ or $N = (L/\lambda)^2$. The total distance traveled backwards and forwards in N steps is $N\lambda \simeq L^2/\lambda$ in diffusing a net distance L. If the total distance L^2/λ exceeds the inelastic mean free path then we can expect inelastic collisions before the electrons cross the discharge.

To look at this issue (of whether electrons cross the discharge before they have an inelastic collision) from the opposite perspective, we can estimate the number of steps N^* of the random walk needed to travel a total distance equal to an inelastic mean free path λ_e^{inel}. The total distance λ_e^{inel} equals N^* multiplied by the step size λ_e^{el} and so $N^* = \lambda_e^{\text{inel}}/\lambda_e^{\text{el}}$. The net distance traveled outward from the starting point before an inelastic collision is then $\lambda_e^* = \sqrt{N^*}\lambda_e^{\text{el}}$ or

$$\lambda_e^* = \left(\lambda_e^{\text{inel}}/\lambda_e^{\text{el}}\right)^{1/2}\lambda_e^{\text{el}} = \left(\lambda_e^{\text{inel}}\lambda_e^{\text{el}}\right)^{1/2}. \tag{4.18}$$

This is the average net distance traveled before a hot electron (which we assume is hot enough to have an inelastic collision) will have an inelastic collision. When $\lambda_e^* > L$ then the ionization is spread throughout the discharge. When $\lambda_e^* < L$ the ionization is localized close to where the heating takes place.

E/N may be large enough within the skin depth beneath the antenna (usually at high neutral pressure) to heat electrons to high energies in a local heating process. However, in all cases we are considering, even at high E/N the heating in an ICP only occurs in a skin depth or so beneath the coil; consequently, in most of the discharge the hot electrons are those hot electrons that diffuse from the region close to the coil.

What is meant by local heating needs more explanation because in the ICP there seem to be degrees of localization. Inductively coupled plasmas are rarely used in a regime where the heating is truly local and where the electron distribution is determined at each point in space by E/N of that point. Three different extents of localization may be distinguished:

1) If the mean free path is large the electrons may be heated in a process that involves multiple passes through the heating region with journeys through the rest of the discharge between each visit to the heating region. This heating process is the least localized – it is entirely nonlocal.

2) If the mean free path is small then the electron may stay in the heating region long enough to have multiple elastic and inelastic collisions. The properties of the heating process will be determined by the local E/N and thus the initial heating of the electrons is local. If these electrons subsequently diffuse into the rest of the discharge and provide ionization throughout the discharge then the rest of the discharge, which is outside the region of high E/N, is heated nonlocally.

3) Finally, if the local electron distribution depends only on the local value of E/N then the heating is entirely local. This may or may not imply that E/N is large enough to produce significant heating and ionization everywhere in the discharge.

Only cases 1) and 2) appear to be relevant to the ICPs used in semiconductor fabrication. See also Refs. [71]–[74].

The other major result of the arguments given so far is that the neutral pressure controls to a great extent where the ionization takes place. The net distance a hot electron travels before having an inelastic collision is $\lambda_e^* = \sqrt{\lambda_e^{\text{inel}}\lambda_e^{\text{el}}}$, provided λ_e^{el}, the elastic *mfp* is considerably less than the system size. This determines how close to the heating region the ionization will occur.

Detailed computational models of ICPs are discussed in Ref. [75]. For a discussion of ion distributions in ICPs, see Ref. [76]; for ICPs used for sputter deposition of metal films see Ref. [77].

In the least localized form of heating, the electrons are heated by a diffusion in energy, a step of the random walk in energy occurring each time they pass the heating region. A similar process occurs in an ECR discharge. A short discussion of this energy diffusion is given next.

4.5 Diffusion in Energy

The "ordinary" diffusion coefficient D is approximately given, in terms of a step size Δ of a random walk and a time step τ of the random walk, as $D \simeq \frac{1}{2}\Delta^2/\tau$. The distanced diffused in a time t is on average $d \simeq \sqrt{2Dt}$. Put differently, the time to diffuse a distance d is about $t \simeq d^2/2D$.

In a similar way, we can define a diffusion coefficient D_E to describe a random walk where the steps are up and down in energy. If the step in energy is Δ_E and the time between steps is τ_E, then the energy diffusion coefficient is

$$D_E \simeq \frac{1}{2}\Delta_E^2/\tau_E. \qquad (4.19)$$

In a time t we can expect the energy to change by about $\delta E \simeq \sqrt{2D_E t}$.

In many plasma reactors, a heating region exists where electrons that enter the region undergo a "kick" in energy. The electrons move around the discharge and occasionally enter the heating region. The size of the kick is assumed to be entirely random, with no correlation between the kicks on successive visits. The mean time between successive visits to the heating region is the mean time between energy steps, τ_E.

It is likely that particles have to diffuse in space, in order to reach the heating zone. The time to diffuse a distance L, the system size, is about $\tau_L \simeq L^2/2D$. Now $D \simeq \lambda^2/2\tau = \frac{1}{2}\lambda v$, where λ is the mean free path, τ is the collision time, and v is the speed; $\tau = \lambda/v$. Then

$$\tau_E = \tau_L \simeq L^2/\lambda v. \qquad (4.20)$$

The energy diffusion coefficient D_E is thus

$$D_E \simeq \frac{1}{2}\Delta_E^2\lambda v/L^2. \qquad (4.21)$$

There is one extra qualification to this: It may be that only some of the particles can reach the heating zone. In this case, if the fraction of particles being heated is f_h, the overall energy diffusion coefficient is f_h times the value given above. However, for those particles being heated, D_E is still given by the above equation.

4.6 Ambipolar Diffusion

The estimates in the previous sections, which are based on a microscopic treatment of diffusion, do not tell the whole story of the charged particle transport, because the electric field in the plasma also drives fluxes. The flux of particles is the net number of particles per second crossing a given area. The flux per unit area Γ is given approximately by

$$\Gamma = -D\nabla n + n\mathbf{v}_d, \tag{4.22}$$

where $\mathbf{v}_d$ is the drift velocity of particles. The charged particles drift along the direction of the electric field $\mathbf{E}$ with velocity $\mathbf{v}_d = \mu\mathbf{E}$, where μ is the mobility. μ is negative for negative particles.

In equilibrium the flux is zero. In one dimension $(\nabla n)_x = \frac{dn}{dx}$ and $E = -\frac{d\Phi}{dx}$, where Φ is the electrostatic potential, and so

$$\Gamma_x = -D\frac{dn}{dx} - n\mu\frac{d\Phi}{dx} = 0. \tag{4.23}$$

This has the solution $n = n_o \exp(-\frac{\mu\Phi}{D})$. Since we expect n to have a Boltzmann form in equilibrium, which means $n = n_o \exp(-\frac{q\Phi}{k_B T})$, where q is the particle charge and T the temperature, we find $\mu/D = q/k_B T$, which gives the Einstein relation

$$D = \frac{k_B T}{q}\mu. \tag{4.24}$$

If q is negative then μ is also negative and so D remains positive.

The derivation of the Einstein relation uses arguments about the fluid in equilibrium to show that the diffusion coefficient for a species of particles is proportional to its temperature multiplied by its mobility – provided a temperature (and D and μ) can be defined for that species. As we shall see, many plasmas are not in equilibrium but only in a steady state where positive and negative charges leave at equal rates. The "ambipolar" diffusion coefficient D_a is an effective diffusion coefficient for the whole plasma, which is derived by requiring the electron and ion fluxes to be equal. D_a is proportional to the electron temperature T_e multiplied by the ion mobility μ_i under the conditions we are usually interested in, when the electrons are much more energetic than the ions.

The calculation of the ambipolar diffusion rate typically starts with the electrons. The electrons are usually so much faster moving than the ions that if there were no electric or magnetic fields the electrons could escape much faster than the ions. The electron and ion fluxes are assumed to be made equal by the electric field, which pulls ions out and holds electrons in, to prevent charge from building up. However, the ion flux is so small from the point of view of the electrons that the term in the expression for the electron flux, which is proportional to the electric field, has to almost entirely cancel the diffusion term, with only a very small remainder, which equals the ion flux. This means that the electrons are nearly in equilibrium, with the electron flux being only a very small fraction of the possible electron flux. As long as we are not taking the electron flux and finding the ion flux by setting the electron flux equal to the ion flux, we can set the electron flux to zero. Then

$$\Gamma_{\text{ex}} = -D_e \frac{dn_e}{dx} + n_e \mu_e \frac{d\Phi}{dx} \simeq 0. \tag{4.25}$$

Since this is a system that is nearly in equilibrium the equation has the usual equilibrium solution:

$$n_e = n_o \exp(e\Phi/kT_e). \tag{4.26}$$

The electron mobility is $-\mu_e$ with the negative sign shown explicitly and $D_e = \frac{k_B T_e}{e}\mu_e$. If we take the natural logarithm of this electron density we find the electrostatic potential Φ in terms of n:

$$\Phi = \frac{k_B T_e}{e} \ln\left(\frac{n_e}{n_o}\right). \tag{4.27}$$

To summarize the argument so far: The electric field makes the electron flux much smaller than it would be if there were no electric field, so that the ions can leave as fast as the electrons. The small electron flux means the electrons are nearly in equilibrium; therefore, the Boltzmann expression for the density applies and this can be used to find the electrostatic potential Φ from the density n_e.

The electric field appears in both the electron and ion fluxes. We can find $E_x = -\frac{d\Phi}{dx}$ by differentiating Φ or by going back to the expression for the electron flux, which gives

$$D_e \frac{dn_e}{dx} = n_e \mu_e \frac{d\Phi}{dx} = -n_e \mu_e E_x. \tag{4.28}$$

Thus

$$E_x = -\frac{D_e}{\mu_e} \frac{1}{n_e} \frac{dn_e}{dx}. $$

Putting this into the ion flux gives

$$\Gamma_i = -D_i \frac{dn_i}{dx} + n_i \mu_i E_x = -D_i \frac{dn_i}{dx} - n_i \mu_i \left(\frac{D_e}{\mu_e}\right) \frac{1}{n_e} \frac{dn_e}{dx}. \tag{4.29}$$

In the bulk of the plasma the electron and ion densities are equal, to achieve quasineutrality, $n_e = n_i$. The ion flux in a quasineutral region is thus

$$\Gamma_i = -D_i \frac{dn_i}{dx} - \mu_i \frac{D_e}{\mu_e} \frac{dn_i}{dx}. \tag{4.30}$$

The ratio $\frac{D_e}{\mu_e}$ appearing here is equal to $k_B T_e/e$, as was derived above, and so

$$\Gamma_i = -\left(D_i + \frac{k_B T_e}{e}\mu_i\right)\frac{dn_i}{dx}. \tag{4.31}$$

Now according to the Einstein relation, $D_i = (k_B T_i/e)\mu_i$. Since we assume the ions are relatively cool ($T_i \ll T_e$) then $D_i \ll (k_B T_e/e)\mu_i$. Neglecting D_i compared to $(k_B T_e/e)\mu_i$, then the ion flux, which equals the electron flux, is

$$\Gamma_i = -(k_B T_e/e)\mu_i \frac{dn_i}{dx}. \tag{4.32}$$

The coefficient that appears in this flux is the ambipolar diffusion coefficient, $D_a = (k_B T_e/e)\mu_i$, as promised. The part of the ion flux we have retained actually came from the mobility term $n_i\mu_i E_x$, which is much bigger than the diffusion term for ions, but the expression for it resembles a diffusion term. The mobility term in the expression for the ion flux is the biggest part of the ion flux because the electric field pulls the ions out of the plasma much faster than the ions could diffuse out unaided.

This picture of the diffusion process is an approximation, with two major flaws: i) The electrons may not have a single temperature T_e and ii) the ion flux is not simply proportional to E_x. These two issues will be described in detail in other applications as we discuss the behavior of the discharges further. The ambipolar diffusion model described here is a very useful tool for estimating the parameters of a discharge analytically, despite these flaws. It is not usually accurate enough to predict densities precisely but if, for instance, we know T_e and n_i then we can estimate the potential from them using the above expressions.

The ambipolar diffusion model given here was one dimensional. No major modification is necessary in two or three dimensions provided the diffusion coefficients and mobilities are the same in all directions. In most of the discharges of interest this is probably true and this model holds approximately in the ICP and rf discharges discussed so far. In the presence of a magnetic field the transport across the field is affected by the magnetic field whereas transport along the field is relatively unaffected by the magnetic field. In the ECR reactor the magnetic field plays a major role in the transport, and the diffusion coefficient is different along and across the field, as we shall see later in Section 4.9.

4.7 Transport and the Plasma Density Profile

The uniformity of processing is largely controlled by the density profile of the plasma. To see how the transport concepts discussed so far control the density profile, an ICP discharge, which sometimes exhibits a hollow density, will be discussed. In a cylindrical chamber enclosing a plasma the peak density is usually found on the cylindrical axis at $r = 0$. If a single-turn antenna is used to power an ICP plasma the power may be deposited off axis. As a result of the heating being away from $r = 0$ it is sometimes possible to get an off-axis density peak at $r \neq 0$. In addition to the antenna providing the power primarily off axis, two other conditions must be met for an off-axis density peak to occur.

The first requirement is that the electrons should not travel too far, between when the antenna's fields heat the electrons and when the electrons turn that energy into ionization. If the electrons only ionize neutral particles, to create ions, in a region close to the antenna then there is a chance that the ion density will peak in that same region.

The second requirement for an off-axis density peak is that the shape of the reactor should allow ions to escape to the chamber walls before they fill in the hollow in the density at $r = 0$. Ions must be able to be lost to the top and bottom faster than they move radially, if the hollow is not to be filled in. Suppose the reactor is a very squat cylinder with a height L, which is much less than its maximum radius R_m. If the ionization happens in a ring close to the coil radius $r = R_c$, this ring will be much closer to the ends of the device than to the center at $r = 0$. If the top and bottom of the reactor are much closer to the ions' point of origin than is $r = 0$, then the ions are much more likely to go to the top or bottom than to $r = 0$ and so the density profile remains hollow at $r = 0$.

To see where the electrons will cause ionization, we must examine the random walk they perform. (We now repeat an argument given earlier.) The net distance an electron goes in a random walk, while having N elastic collisions with neutrals, with the mean free path for elastic collisions being λ_e^{el}, is roughly $\sqrt{N}\lambda_e^{el}$. The total distance it travels is $N\lambda_e^{el}$ and after some particular number of elastic collisions N^* the total distance $N^*\lambda_e^{el}$ becomes comparable to the mean free path for ionization λ_e^{ion}. After N^* steps and because the total distance traveled is $N^*\lambda_e^{el} = \lambda_e^{ion}$, an ionization event is likely. The number N^* is found from $N^*\lambda_e^{el} = \lambda_e^{ion}$ to be $N^* = \lambda_e^{ion}/\lambda_e^{el}$. Because the distance traveled away from where the electron started in N steps is $\sqrt{N}\lambda_e^{el}$ the net distance λ_e^* traveled before ionization is about $\sqrt{N^*}\lambda_e^{el}$ or $\lambda_e^* = \sqrt{\lambda_e^{el}\lambda_e^{ion}}$. If this length λ_e^* is much less than the coil radius R_c then the ionization will happen close to the coil.

Suppose $\lambda_e^{ion} \sim 10\lambda_e^{el}$ and that the coil radius $R_c \sim 15$ cm. For $\lambda_e^* \ll R_c$ we must have $\sqrt{10}\lambda_e^{el} \ll 15$ cm or $\lambda_e^{el} \ll 5$ cm. Since $\lambda_e^{el} = 1/n\sigma$ and if $\sigma \sim 10^{-20}$ m^2 = 10^{-16} cm^2 then for $\lambda_e^{el} \sim 1$ cm we must have a neutral density $n = 10^{16}$ cm^{-3}. If $\sigma \sim 10^{-15}$ cm^2 instead then $n = 10^{15}$ cm^{-3} when $\lambda_e^{el} = 1$ cm. A pressure of 1 mTorr corresponds to 3×10^{13} cm^{-3}; so $n = 10^{15}$ cm^{-3} is about 30 mTorr.

If the height of the reactor L is much bigger than its radius R_m then the diffusion is essentially radial. We now examine the case of purely radial diffusion, to show what can be expected in this limit. Particles go to the wall at $r = R_m$ or fill in the density $r = 0$ far faster than they go to the top or bottom. Mathematically, the steady-state continuity equation for a diffusion process in one dimension r is

$$\nabla \cdot \mathbf{\Gamma} = \frac{1}{r}\frac{\partial}{\partial r}r\left(-D\frac{\partial n}{\partial r}\right) = S \qquad (4.33)$$

since

$$\Gamma_r = -D\frac{\partial n}{\partial r}. \qquad (4.34)$$

In the plasma the diffusion process is ambipolar diffusion. Ions are actually pulled out by the electric field, but this equation still holds, at least at higher pressures, with $D = D_a$.

We can integrate this once over r to get the radial flux per unit area:

$$\Gamma_r = -D\frac{\partial n}{\partial r} = \frac{1}{r}\int_o^r r'S dr'. \qquad (4.35)$$

This integral has a very simple explanation. The net flux passing through a radius r (in steady state) consists of all the particles produced per second between the center at a radius of zero and the radius in question. No net flux of particles passes through $r = 0$ because of symmetry. If we integrate the source per unit volume S over unit length in the z direction (which has no effect at all since $S(r)$ depends only on r) and over the length associated with the cylindrical angle ϕ, which multiplies $S(r)$ by $2\pi r$, the production rate per unit length in z and per unit radius is $2\pi r S(r)$. The total flux leaving radially (per unit in z) is the integral over radius of this production rate,

$$\Gamma_{TOT} = \int_0^r 2\pi r' S(r')dr'. \qquad (4.36)$$

The flux per unit area is the total flux, divided by the area the flux passes through. The area A is the area of the outer surface of a cylinder, which consists of unit length in z multiplied by a circumference of $2\pi r$: $A = 2\pi r$ and $\Gamma_r = \Gamma_{TOT}/A$ or

$$\Gamma_r = \frac{1}{2\pi r}\int_0^r 2\pi r' S(r')dr' = \frac{1}{r}\int_o^r r'S dr'. \qquad (4.37)$$

This flux per unit area is equal to $-D\frac{\partial n}{\partial r}$, which is the same as the diffusion equation found above.

If no particles are produced within a radius r then no net flux passes through that radius. If the flux $\Gamma_r = 0$ then $\frac{\partial n}{\partial r} = 0$ and n is flat. When S becomes positive Γ_r becomes positive – there is an outward flux. If $\Gamma_r > 0$ this means $\frac{\partial n}{\partial r} < 0$ and so the density goes down as r increases.

This analysis shows that if diffusion is one dimensional, and no flux goes through the boundary on one side, the density is maximum at the side with no net flux. Even if S is only positive away from $r = 0$, the density is still biggest at $r = 0$ in a one-dimensional case. (Negative S is not considered, here.)

In two dimensions we need to estimate the importance of the second dimension, z. We can estimate the time to diffuse to the ends by going a distance $L/2$ and compare it to the time to diffuse from $r = R_c$ to $r = 0$, to see how large we can expect the difference made by loss to the ends to be. In time t the mean distance diffused is about $\sqrt{2Dt}$. Thus to diffuse about $L/2$ to the end will take a time

$$t = (L/2)^2/2D = L^2/8D. \qquad (4.38)$$

To diffuse a distance R_c will take a time $t = R_c^2/2D$. If $R_c \gg L$ then the particles will reach the ends much faster than they will reach $r = 0$. The combined implication of these estimates is that if the neutral pressure is a few tens of mTorr or more (to keep the ionizing electrons close to the coil) and the chamber has a height much less than its radius to prevent the density filling in at $r = 0$, then an off-axis density peak can form.

4.8 The Plasma Potential

Next we will examine the potential set up as a result of the electron behavior in the ICP. First we can predict that the electrons must be effectively confined in the discharge to enable them to ionize neutrals. If the kinetic energy an electron must have to ionize a neutral is ε^*, then the electrostatic potential at the wall must usually be high enough to keep particles with this energy in the discharge. At higher pressures the collisions might slow electron diffusion down enough to allow electrons to ionize neutrals before they reached the wall, but this is not so likely at the mTorr pressures used in the ICP. The wall potential in volts must therefore be bigger than ε^* in electron volts, to confine the ionizing electrons.

If the ionization takes place in a multistep process, for instance one where the electrons excite neutrals and the excited neutrals ionize each other, then the energy the electrons must have to initiate this process is lower than ε^*. If multistep ionization is dominant then the confining potential can be lower than for a case dominated by direct ionization.

4.9 Analytic Models of Electron Cyclotron Resonance Discharges

ECR discharges, in use for plasma etching [63] or deposition in microelectronics applications, usually use microwaves at a frequency of $f = 2.45$ GHz to power the discharge. The electrons must gyrate at (nearly) the same frequency to be resonantly heated. The gyrofrequency in the magnetic field is $\omega_c = eB/m$; so if we set $f = \omega_c/2\pi = 2.45 \times 10^9$ s^{-1} then $B = 2.45 \times 10^9(2\pi m/e) = 8.75 \times 10^{-2}$T or 875 gauss for resonance. This magnetic field makes the electron gyroradius (the radius of the circle the electron orbit makes in a magnetic field) very small while the ion gyroradius is typically about 1 cm. The effect of the difference in gyroradii on the charged particle transport is perhaps the major issue that distinguishes the ECR source from other reactors. In this section the focus will be on the transport of electrons and ions in the presence of this magnetic field and the way that electric charge build-up on the chamber walls might influence that transport.

The gyroradius is given by $r_\perp = v_\perp/\omega_c$, where $v_\perp$ is the particle speed in the plane perpendicular to the magnetic field line. For electrons $v_{\perp e} \sim 10^6$ ms^{-1} and $\omega_{ce} \sim 2\pi(2.45 \times 10^9)$ radians/s; so $r_{\perp e} \sim 10^{-4}$ m. For the ions the gyrofrequency is lower because of their higher mass, by a factor of the order of the mass ratio. Therefore $\omega_{ci} \sim 10^5$ radians/s. The ion perpendicular velocity might be (at most) $v_{\perp i} \sim 300$ m/s, making the ion gyroradius $r_{\perp i} \sim 3 \times 10^{-3}$ to 10^{-2} m or 0.3 to 1 cm.

The diffusion across the field lines occurs when particles that are gyrating about one field line have a collision. The collision makes the particles "forget" which line they are orbiting. The new line they go into orbit around can be as much as a gyroradius (or two) from the first line; hence the step size of the diffusion is about

$r_\perp$ and the time between steps is the collision time τ_c. The perpendicular diffusion coefficient is $D_\perp \simeq \frac{1}{2} r_\perp^2 / \tau_c$.

The alternative form of the diffusion coefficient, $D \simeq \frac{1}{2} v \lambda$ (where v is the speed and λ is the step size, which is usually a mean free path) is not appropriate for diffusion of charged particles across a strong magnetic field – that is, a magnetic field which makes the charged particles gyrate around the field lines with $r_\perp \ll L$, the system size. The speed v in this expression must be equal to the step size divided by the time per step, for the different forms of diffusion coefficient to be equivalent. Since the particle may make many gyroorbits in the collision time τ_c, the step size $r_\perp$ and the time per step are not related to each other by the speed. The time τ_c itself is still related to the mean free path λ by $\tau_c = \frac{\lambda}{v}$ but λ is not the step size in this diffusive process.

The diffusion coefficient can be written in terms of λ starting from $r_\perp^2 / 2\tau_c$, by eliminating $\tau_c = \lambda / v$. $D_\perp$ becomes

$$D_\perp \simeq r_\perp^2 v / 2\lambda. \tag{4.39}$$

So if we assumed $\lambda \sim 5$ cm for both species we would have $D_{\perp e} \sim (10^{-4})^2 10^6 / 0.1 = 10^{-1}$ m^2/s and $D_{\perp i} \sim (10^{-2})^2 10^3 / 0.1 \simeq 1$ m^2/s. The diffusion coefficients along the magnetic field, $D_\parallel$, are not affected by B. We have $D_{\parallel e} \sim \frac{1}{2} v \lambda \simeq \frac{1}{2}(10^6)(0.05) = 2.5 \times 10^4$ m^{2}s^{-1} and $D_{\parallel i} \sim \frac{1}{2}(10^3)(0.05) = 25$ m^{2}s^{-1}. From these diffusion coefficients we can now estimate the characteristic times (ignoring electric field effects at first).

The time $t_\perp$ to diffuse a distance L of about 5 cm to the wall is roughly given by $L = \sqrt{2 D_\perp t_\perp}$ or $t_\perp = L^2 / 2 D_\perp$. For electrons $t_{\perp e} \simeq (5 \times 10^{-2})^2 / (2 \times 10^{-1})$. So $t_{\perp e} \simeq 10^{-2}$ s. Similarly $t_{\perp i} \simeq (5 \times 10^{-2})^2 / (2) \simeq 10^{-3}$ s. To diffuse ten times as far along the field line to the end of the tube takes an electron about $t_{\parallel e} \simeq (0.5)^2 / (5 \times 10^4) = 5 \times 10^{-6}$ s. An ion will take approximately $t_{\parallel i} \simeq (0.5)^2 / (50) \simeq 5 \times 10^{-3}$ s.

These estimates of the times show that electrons could diffuse down the magnetic field a distance of $L = 0.5$ m in about 5×10^{-6} s. To get to the radial wall directly takes electrons a much longer time of 10^{-2} s. The electrons still need to overcome a potential barrier in the sheath before they get to the wall in either case; thus these times are underestimates of the actual times taken because not all electrons can escape over the potential barrier. However, for electrons that are energetic enough to escape, the relative sizes of these times are probably roughly correct. So electrons will prefer to escape to the ends of the system. Ions can escape to the radial wall in 10^{-3} s and will take about five times as long to reach the ends. Therefore, ions are most likely to go to the walls.

If the walls and the ends conduct electricity, the electrons that hit the ends are conducted through the sidewalls to the locations on the walls where the ions strike. The electrons can then neutralize the ions, since electrons and ions leave the plasma at equal rates in steady state if the ions are singly charged. (If the ions have p positive charges each, then p times as many electrons leave per second as ions in steady state.) If the walls are insulators the situation is much more complex since charge will build up on the walls until an electric field is set up to prevent the charge building up any further.

A review of recent results on ECR discharges is given in Ref. [78]. Five representative ECR designs and performance figures of merit derived from available experiments are discussed.

4.10 The Electron Distribution Function

4.10.1 Introduction

The distribution function f is the number of particles per unit volume in phase space. The velocity distribution is the number per unit of the components of the velocity and per unit volume. So, for instance, the number in $(dx, dy, dz, dv_x, dv_y, dv_z)$ at (x, y, z, v_x, v_y, v_z) is $f(x, y, z, v_x, v_y, v_z)dxdydzdv_xdv_ydv_z$. This will sometimes be written as $f(\mathbf{x}, \mathbf{v})d\mathbf{x}d\mathbf{v}$. The energy distribution multiplied by $d\varepsilon dxdydz$ gives the number in $d\varepsilon dxdydz$ (where ε is the kinetic energy) and so on.

The distribution function is constant along a particle trajectory in phase space, in the absence of collisions:

$$\frac{df}{dt} = 0, \tag{4.40}$$

where the derivative is at the particle position $\mathbf{x}$, $\mathbf{v}$. An alternative derivation of the "kinetic equation" uses a statement that f is conserved by the flow in phase space; it is a generalization of the continuity equation to include the extra independent variables in phase space, in addition to the spatial coordinates. Collisions add an extra term on the right-hand side: $C(f)$.

The density is obtained from the velocity distribution f by integrating over the velocity variables:

$$n = \int f(\mathbf{x}, \mathbf{v})dv_xdv_ydv_z. \tag{4.41}$$

Since $\varepsilon = \frac{1}{2}mv^2$, if we integrate the velocity distribution over angle in velocity space and rewrite the element of volume in velocity space as $2\pi v^2 dv$, and then switch to kinetic energy as the variable we integrate over, this volume element becomes $2\pi(2\varepsilon/m^3)^{1/2}d\varepsilon$. When we instead integrate the energy distribution $f(\varepsilon)$ over energy, to obtain the density, the integral is $\int f(\varepsilon)d\varepsilon$. The above result for the volume element, however, shows that if we integrate the velocity distribution over energy, instead of speed, the integral that gives the density will have an extra factor of $\varepsilon^{1/2}$.

The use of the distribution function to calculate rates of processes that are driven by electrons is discussed in Chapter 5.

4.10.2 Physical Processes and the Electron Distribution

The electron distribution function, or EDF, describes the number of electrons to be found in each range of energies. The chemical processes taking place in a plasma are driven by the electrons, and so knowledge of the electron distribution function is necessary for understanding plasma chemistry. How the electrons are distributed in energy is usually determined by the competition between the heating process and the collisions that remove the electrons' energy. The energy input is typically by means of externally applied electric fields. The role of the collisions and of the heating mechanisms in determining the energy distribution will be examined in this section.

The main classes of electron collisions we have to consider are:

1) Elastic collisions with neutrals. The neutrals are so much more massive than the electrons that the neutrals recoil only slightly when an electron hits them, and so they pick up a very small fraction of the electron's energy. For electrons that are trapped in a potential well this slow elastic cooling will tend to push their distribution toward being a Maxwellian at the temperature of the neutral gas.

2) Inelastic collisions with neutrals. When an electron's kinetic energy is above the threshold in kinetic energy, an inelastic process can occur. The inelastic process happens when the electron hits the neutral and gives up what is typically a substantial amount of its kinetic energy to the neutral, and the energy is used to excite the neutral. If the neutral is a single atom, the inelastic threshold is high ($\sim 10\,\mathrm{eV}$) because the excitation involves a change in the electronic energy level. In diatomic molecules, energy can be stored in vibrational and rotational states, for which the threshold is a fraction of an electron volt. (If diatomic molecules are present, the inelastic threshold is very low and nearly all the electrons can have inelastic collisions.) If the neutral has its electronic state changed, this may lead directly to ionization; or if the new electronic state is a long-lived neutral the energy may be used when two such excited atoms collide, to ionize one of them. Otherwise the energy is likely to be lost from the discharge when the neutral hits a wall and gives up its excitation energy. Energy put into vibration or rotation is usually given to other neutrals that collide with the excited neutral. The other neutrals take the energy away as translational energy – that is, it is used in heating up the neutrals. This process tends to be very efficient.

3) Coulomb collisions with other electrons. In some circumstances the cooler electrons, which are well below the inelastic threshold, may nevertheless be losing energy to elastic cooling, having no way of being heated except by Coulomb collisions with more energetic electrons [79].

The electric (and sometimes magnetic) fields also play a crucial role in determining the electron distribution. They heat the electrons, and the static electric field holds the electrons inside the plasma. The confining field has to ensure that overall, in steady state, electrons leave at the same rate as ions (if the ions are singly charged) and that rate at which electrons (and ions) leave is equal to the overall rate at which electrons (and ions) are produced.

One of the critical features of the heating mechanism is the group of electrons it affects. Some heating processes only happen at the edge of the discharge, whereas some extend throughout the plasma. Only electrons that are already energetic can get from the middle of the plasma to the edge. Consequently, the heating of electrons that takes place at the edge puts the energy into the high-energy group. This may leave most of the electrons cold, since they will not be able to get into the region where heating happens. The hot group of electrons are then the only electrons that get heated effectively, and thus a group of electrons that is only loosely connected with the bulk of the electron distribution will be responsible for all the ionization. However, if the heating field reaches all the way into the plasma it is likely to heat all the electrons, enabling them to all gain energy and participate in causing ionization.

To summarize part of this problem of determining the electron distribution, we have to consider the interaction between several effects, namely i) heating of electrons (usually by externally supplied electric fields), ii) loss of energy from electrons to neutral particles, and iii) confinement of the vast majority of electrons, (which are less energetic) inside the plasma, and of the hot electrons long enough for them to produce sufficient ionization to sustain the discharge. The confinement is usually by the electrostatic field and is possibly assisted by or even primarily by elastic collisions, which strongly affect confinement at high neutral pressure. The confinement of cool electrons is accompanied by elastic cooling and by loss of the most energetic electrons to the walls, or to ionization. So since the cool electrons do not, in practice, cool down to a very low temperature, they must be heated in some way. Redistribution of energy between electrons by Coulomb collisions may sometimes be important.

Deviation from Maxwellian

The electron distribution is frequently assumed to have a "Maxwellian" form, which is the equilibrium shape of the distribution function. In plasmas used in semiconductor processing this is rarely true – the EDF is quite unlikely to be Maxwellian, at least that part of the distribution corresponding to energies above the inelastic threshold. In monatomic gases, an electron kinetic energy $\gtrsim 10$ eV is required so that electrons could have inelastic collisions (or escape to the walls). As is the case for many approximations that are made to estimate plasma behavior, it is important to distinguish when the approximation (in this case, of a Maxwellian EDF) can be used to represent the appropriate physical situation and when it cannot. Some discharges such as the ICP have a more Maxwellian EDF than others. However, even in the ICP where it is useful to assume a Maxwellian EDF to calculate some quantities, it would nevertheless be very misleading to use a Maxwellian in some other contexts.

The high energy part or "tail" of the EDF is particularly likely to deviate from a Maxwellian. The tail electrons can often escape to the walls of the chamber and they have enough kinetic energy to have inelastic collisions in which they lose large amounts of kinetic energy. Both of these processes remove electrons from the tail. The number in the tail consequently tends to be lower than if the tail had the equilibrium shape that matched the shape of the rest of the electron distribution. If the rest of the electrons, which are less energetic than the tail electrons, have a Maxwellian shape, an extrapolation of that Maxwellian to the energies of the tail will usually overestimate the number in the tail. In that case, using a Maxwellian assumption based on the distribution of cooler electrons will overestimate the rate of any process that depends on the tail electrons, such as ionization, excitation, or loss of electrons to the walls. If the power being put into inelastic processes was known, then assuming a Maxwellian would lead to an underestimate of the temperature T_e of the bulk of the electrons. This leads to an underestimate of the plasma potential, since $\Phi \sim T_e$.

Coulomb collisions between electrons can transfer energy to make the EDF more Maxwellian but this is usually a slow process and has little chance of affecting the tail. In the bulk of the EDF, which consists of the cooler electrons, the Coulomb collisions can sometimes provide an equilibrium between the electrons at some temperature T_e. Alternatively, local heating of the electrons may result in them having a temperature T_e that depends on E/N. Properties of the plasma that depend on the bulk of the electrons can in some cases be estimated using a Maxwellian EDF, even though properties that depend on the tail cannot be found reliably in this way. The best example of this is the overall density of the electrons, which depends on the cooler bulk of the EDF since only a very small fraction of the electrons reside in the tail. In other words the Boltzmann density distribution $n = n_o \exp(e\Phi/k_B T_e)$ might work well if the bulk of the EDF has a temperature T_e.

As mentioned in this section, the ICP often has a Maxwellian EDF in the bulk. The high density of electrons makes the Coulomb collision frequency ν_{ee} relatively high compared to ν_{ee} in many processing discharges, and the collisions probably cause "equilibration" among the bulk electrons. The ECR discharge has a similarly high electron density. However, in the ECR discharge the microwave power is supplied directly to the bulk electrons and the gentle effect of Coulomb collisions cannot compete with this to shape the EDF. In the ICP the heating is not applied directly to the bulk and in fact the bulk electrons are primarily heated by Coulomb collisions with hotter electrons. Thus the part of the EDF that is directly heated in these discharges seems to be a factor in deciding whether the bulk is in equilibrium or not. Electrons in ECR discharges are observed to often have a biMaxwellian distribution, which means that the electron distribution can be approximated as the sum of two Maxwellians at different temperatures. In ICPs the EDF often appears to be a single Maxwellian, at least the lower energy group. In all cases the tail is very hard to observe because there are relatively few electrons in it and so measurements of the tail are not generally available. In rf capacitive discharges the bulk electron distribution may have an equilibrium form at a temperature T_e if the heating is local. The temperature T_e could then be found from a table as a function of E/N if the data are available for the neutral gas employed. If the heating is nonlocal the EDF will very likely not have an equilibrium shape.

4.10.3 Electron Distribution Function Depending on Total Energy

In a low-pressure plasma the case can often be made that the electron distribution function is approximately a function of the total energy of the particles. In the absence of collisions, the particles move around throughout the energetically accessible part of the discharge. As they move, they have the same total energy (by conservation of energy) and the same distribution function. The distribution function f is constant because the kinetic equation can be written as $\frac{df}{dt} = 0$ along the particle trajectories, in the absence of collisions.

Elastic collisions of electrons change the electrons' direction, but since the collisions are with much more massive neutrals, the energy is changed very little. In particular, if the distribution is nearly isotropic anyway, elastic collisions with neutrals have little effect on it. So in the range of total energies where particles never pick up enough kinetic energy to have inelastic collisions, we expect the electron distribution to be nearly a function of the total energy. (If inelastic collisions are possible but they remove little energy, for instance when they excite vibrational or rotational states in a diatomic molecule, we may still be able to ignore the energy loss, if these inelastic collisions are infrequent.) At energies where inelastic collisions have a significant effect, or if the electrons in question are strongly heated irreversibly, for example by an external rf field, the distribution will not simply be a function of total energy.

The usual form of the distribution is an exponential decay with energy. As the energy increases the slope may change, especially as the inelastic threshold is approached. Suppose that the velocity distribution f is similar to a Maxwellian, but that f varies with total energy $\varepsilon_T \equiv \varepsilon - e\Phi$ as

$$ f = f_o \exp^{(-a\varepsilon_T - b\varepsilon_T^2 - c\varepsilon_T^3)}. \tag{4.42} $$

It is interesting to relate this to the electron temperature. The fluid equation for the electron flux per unit area reads

$$ \mathbf{\Gamma} = -n_e \mu_e \mathbf{E} - D_e \nabla n_e, \tag{4.43} $$

where the electron mobility is $-\mu_e$. This flux is usually roughly zero. (Compared to the size of each of the terms separately, the two terms usually nearly cancel).

If the number density of electrons is $n_e = \int_0^\infty \varepsilon^{1/2} f \, d\varepsilon$ then if f is a function of ε_T we find

$$ \nabla n_e = \int_0^\infty \varepsilon^{1/2} \nabla f \, d\varepsilon = \left[\int_0^\infty \varepsilon^{1/2} \frac{\partial f}{\partial \varepsilon_T} d\varepsilon \right] \nabla \varepsilon_T = n' \nabla \varepsilon_T. \tag{4.44} $$

But since $\nabla \varepsilon_T = -e\nabla\Phi = e\mathbf{E}$, $\mathbf{\Gamma} = -(n_e \mu_e + D_e n' e)\mathbf{E} \simeq 0$. This means that $\frac{D_e e}{\mu_e k_B}$, which we shall use for now to define the temperature, $T_e \equiv \frac{D_e e}{\mu_e k_B}$, is given by $\frac{D_e e}{\mu_e} \equiv k_B T_e = -n/n'$.

Now n and n' are both dominated by the lower limit, $\varepsilon = 0$, of the integrals that give them in terms of f, and so

$$ \frac{n}{n'} \simeq f \Big|_{\varepsilon=0} \Big/ \frac{\partial f}{\partial \varepsilon_T} \Big|_{\varepsilon=0}. \tag{4.45} $$

Given our assumption that $f = f_o \exp^{(-a\varepsilon_T - b\varepsilon_T^2 - c\varepsilon_T^3)}$, then $\frac{\partial f}{\partial \varepsilon_T} = -(a + 2b\varepsilon_T + 3c\varepsilon_T^2)f$, and at $\varepsilon = 0$ where $\varepsilon_T = -e\Phi$ for electrons,

$$ k_B T_e \simeq [a + 2b(-e\Phi) + 3c(-e\Phi)^2]^{-1}. \tag{4.46} $$

If we let $T_o = (k_B a)^{-1}$, then $T_e = T_o/(1 + b_1\Phi + c_1\Phi^2)$, where $b_1 = -2be/a$ and $c_1 = 3ce^2/a$.

This result for the temperature is interesting, partly because if we write the flux per unit area

$$ \mathbf{\Gamma} = -n_e \mu_e \mathbf{E} - D_e \nabla n_e \simeq 0 \tag{4.47} $$

then $n_e \mu_e \nabla\Phi = D_e \nabla n_e$ and

$$ \nabla\Phi = \frac{D_e}{\mu_e} \frac{1}{n_e} \nabla n_e = \frac{k_B T_e}{e n_e} \nabla n_e. \tag{4.48} $$

But if $T_e = T_o/(1 + b_1\Phi + c_1\Phi^2)$, then

$$(1 + b_1\Phi + c_1\Phi^2)\nabla\Phi = \frac{k_B T_o}{en_e}\nabla n_e, \qquad (4.49)$$

which means

$$\nabla\left[\Phi + \frac{b_1}{2}\Phi^2 + \frac{c_1}{3}\Phi^3\right] = \frac{k_B T_o}{en_e}\nabla n_e. \qquad (4.50)$$

This can be solved to obtain

$$\frac{k_B T_o}{e}\ln\frac{n_e}{n_o} = \Phi + \frac{b_1}{2}\Phi^2 + \frac{c_1}{3}\Phi^3, \qquad (4.51)$$

which in turn means

$$n_e = n_o\exp\left[e\left(\Phi + \frac{b_1}{2}\Phi^2 + \frac{c_1}{3}\Phi^3\right)\bigg/k_B T_o\right]. \qquad (4.52)$$

This reasoning also means that since n_e and T_e can be written as functions of Φ, then for instance T_e and Φ can both be written as functions of n_e.

4.10.4 Excitation and Ionization Rates

If the electrostatic potential was actually constant inside the plasma, and if electrons could travel from the heating region and bounce backwards and forwards throughout the entire plasma volume before they lost energy to inelastic collisions, then the electron distribution might be uniform in space. Inside the quasineutral plasma, the potential does vary, by amounts of the order of $k_B T_e/e$. This variation can have a significant effect on inelastic rates.

a) $\lambda_e^* \gg L$

The distribution function decreases very rapidly with increasing energy. Suppose that in the center of the plasma there is a density N^t of electrons at the threshold in kinetic energy for an inelastic process. The density of electrons with kinetic energy $(k_B T_e)$ greater than the threshold energy is almost always a lot less; say $f^t N^t$, where $f^t \ll 1$. If the inelastic mean free path is large, then at a point where the electron potential energy is $(k_B T_e)$ greater, the density of electrons at threshold is about $f^t N^t$. This in turn means that the rate for the inelastic process is smaller by a factor of about f^t at that point.

b) $\lambda_e^* \ll L$

There is another reason why the inelastic rates vary in space when λ_e^*, the effective length for electron energy loss, is smaller than L. This is that hot electrons are produced in one region, and they lose their energy by having inelastic collisions before they cross the discharge. Further, hot electrons that are produced in one location but that are below the threshold in energy may be heated to above threshold, if that point is at a "high" potential for electrons. They accomplish this by falling through the potential gradient and gaining enough kinetic energy to reach the threshold energy. They can then undergo inelastic collisions, and if $\lambda_e^* \ll L$, these inelastic collisions will be close to the point where they reach the threshold energy. (If $\lambda_e^* \gg L$, they will have inelastic collisions throughout the entire region where their kinetic energy is above the threshold.)

In both cases the rates varied because of the potential Φ. It was assumed in case a) that the reactor had a low enough neutral density so that the inelastic mean free path λ_e^{inel} was large compared to the system size L. The elastic mean free path λ_e^{el} is usually small compared to L.

Returning to case b), if $\lambda_e^{\text{el}} \ll L$ the elastic collisions are probably frequent enough to make the electron distribution isotropic. (As was discussed earlier, the net distance traveled before an inelastic collision is much smaller than λ_e^{inel}, because of the frequent changes in direction caused by elastic collisions.)

The part of the distribution we are usually most interested in and the one that drives ionizing collisions has a kinetic energy much greater than $k_B T_e$. (If the inelastic threshold is very low this will not be true.) Because of this, the speed of these particles will not change much if their kinetic energy changes by about $k_B T_e$. The inelastic rates depend on particle kinetic energy and on numbers – but in this case, the change in kinetic energy is small while the change in numbers is large. To find the variation in inelastic rates, we primarily need to know how the numbers of electrons vary with energy. To do this, we need to find the distribution function – a genuinely difficult task in many cases. This is why we simply estimated the way the distribution varies with energy. To do better than this, we probably need a numerical calculation.

Now consider case b) again. Suppose that the net distance traveled before an inelastic collision is small compared to the system size. The total distance traveled before the inelastic collision is λ_e^{inel}, but since the particle travels in steps of length λ_e^{el}, the elastic mean free path, the number of elastic collisions before an ionizing collision is roughly $N^* = \lambda_e^{\text{inel}}/\lambda_e^{\text{el}}$. The average distance diffused in N steps of length λ is $\sqrt{N}\lambda$, which in this case gives an average length for energy loss of $\lambda_e^* = \sqrt{\lambda_e^{\text{inel}}\lambda_e^{\text{el}}}$. The ionization rate will vary as $\frac{1}{r^2}e^{-r/\lambda_e^*}$ with distance r from a point source of energetic electrons. For a line source, the r^2 is replaced by r.

4.10.5 Higher Pressure Discharges

In the context of plasma etching, a "high-pressure" discharge might be at a pressure $p \gtrsim 100$ mTorr. In these higher pressure cases, the electron distribution is likely to be determined by local conditions at each point in the discharge. At low pressure, electrons can travel across the discharge with little chance of inelastic collisions, and so the energy distribution is determined by the average conditions in the accessible volume. At high pressure, the electrons lose energy to inelastic collisions before they can go far. Thus they rapidly "forget" conditions at one location when they move somewhere else and consequently the distribution of energies at any point only depends on local conditions at that point.

To describe a discharge where the local field determines the electron distribution, one can use swarm data, as mentioned previously. The mobility is a meaningful quantity under these local conditions. If it is known, and the ionization and other inelastic rates per electron are known as functions of the electric field, then a useful fluid model of the discharge can be set up. Alternatively, the computational method described in Section 4.11 can be used since it is very successful at predicting swarm data, provided the appropriate cross sections are known.

The density of electrons is found from the continuity equation, where the ionization rate appears. The ionization rate (S) depends on the electric field and the electron density. The electric field is usually calculated from Poisson's equation, but in the bulk plasma it is controlled by continuity of current since the current density $\mathbf{J} = n\mu_e\mathbf{E}$. Thus μ_e determines $\mathbf{E}$, which determines S, and from S the density is found.

In these "local" conditions the ionization rate $S = n_e f(E)$, where $f(E)$ is a function of the electric field, which rises roughly exponentially with E in some ranges of parameters. Suppose we write $f(E) = A \exp(E/E_c)$. The electron current density is $J_e = n_e \mu_e E$. Then $E = J_e/n_e\mu_e$ and $f(E) = A\exp(J_e/n_e\mu_e E_c)$ so the ionization rate is $S = n_e A \exp(J_e/n_e\mu_e E_c)$. It is interesting to examine the shape of S predicted by this expression. In particular, the maximum ionization rate occurs at the point in the discharge where $\frac{\partial S}{\partial n_e} = 0$. Now

$$\frac{\partial S}{\partial n_e} = A\exp(J_e/n_e\mu_e E_c)\left[1 - \frac{J_e}{n_e\mu_e E_c}\right]. \tag{4.53}$$

This equals zero where $n_e = n_e^S \equiv J_e/\mu_e E_c$. Since $J_e = n_e\mu_e E$, $n_e^S = \frac{n_e E}{E_c}$. Equivalently, the ionization is maximum where $E = E_c$. Now in practice the field E_c is probably large compared to the typical field in the bulk of the discharge. In this case, the maximum ionization rate is not in the center of the discharge but in the lower density regions, near the sheaths. If E_c is small, the density n_e^S is reached only near where E crosses zero.

4.10.6 Inelastic "Holes" in the Electron Distribution Function

Inelastic collisions remove electrons from one range of energies and put them back in another (usually lower) range of energies. Suppose that the cross section σ^{inel} for a particular inelastic collision was very large. In that case, the inelastic collision might nearly empty the electron distribution, in the range of energies where the cross section was large. The sort of analytic estimates we usually make of the inelastic rates tend to assume the electron distribution has a smooth shape that is close to being Maxwellian. If there are "holes" (caused by inelastic processes) in the distribution, then the inelastic rates must be calculated more carefully. It is increasingly expected that even single-component plasmas do not have Maxwellian electron distributions [80, 81]. The role of inelastic collisions in this is not fully clarified yet.

The calculation will require a knowledge of the rate R_h at which electrons are being fed in the "hole". Some fraction α_{in} of those electrons will lead to inelastic events. The rate of inelastic events will be $\alpha_{in} R_h$. The fraction α_{in} that leads to inelastic events can be found from the frequency v_e^{inel} of inelastic events and the frequency of leaving the hole by any other method.

The frequency with which electrons, having energy above the inelastic threshold, have inelastic collisions is

$$v_e^{inel} = n\sigma^{inel} v_e = v_e / \lambda_e^{inel}, \tag{4.54}$$

where n is the neutral number density and v_e will be about 10^6 m/s. The inelastic cross section $\sigma^{inel} \leq 10^{-20}$ m^2. The neutral density n could range from 10^{19} to 10^{22} m^{-3} and beyond, in which case the mean free path λ_e^{inel} runs from 10 m to 0.01 m, and v_e^{inel} is from 10^5 to 10^8 s^{-1}. If $\lambda_e^{inel} = 0.01$ m, it is hard to imagine another process that could compete with the inelastic events; certainly, electrons go much further than this to reach the wall. Because the frequency $v_e^{inel} = 10^8$ s^{-1} is also much higher than any other characteristic frequency for energy gain or loss, $\alpha_{in} \simeq 1$. At the other extreme, if $\lambda_e^{inel} = 10$ m, electrons could bounce off the sheath many times before an ionization, and the bouncing could fill in the hole.

Suppose that the electron distribution was Maxwellian, with a number density of n_e. Then the fraction of these particles that could cause an inelastic event is probably between 0.1 and 10^{-3}, depending on the electron temperature T_e. If the inelastic threshold E_{th} for a chemical reaction is about $2T_e$ (in suitable units) then the number density of electrons available might be about $n_e/10$. The frequency v_e^{inel} we estimated above is the rate at which each electron with energy above the inelastic threshold collides inelastically. The total rate per unit volume of inelastic events would then be $R^{inel} = v_e^{inel} n_e / 10$. The highest value we estimated of $v_e^{inel} \simeq 10^8$ s^{-1} would thus give $R^{inel} \sim 10^7 n_e$. For a typical electron density in an ECR or ICP device of $n_e \sim 10^{18}$ m^{-3}, this is $R^{inel} \sim 10^{25}$ m^{-3}s^{-1}. The energy input per event is about $E_{th} \sim 5$ eV, which corresponds to about 8×10^{-19} J, giving the power input per unit volume of $P_{in} = R^{inel} E_{th} \sim 10^7$ Wm^{-3}. The volume V_r of a plasma reactor typically corresponds to that of a cylinder, which is about a third of a meter in each direction, and so $V_r \sim 3 \times 10^{-2}$ m^3. The power input $P_{in} \sim 3 \times 10^5$ W found in this way is many orders of magnitude too big to be correct. Thus for this value of v_e^{inel} the inelastic collisions would easily "burn a hole" in the electron distribution. Once the hole was present, the rate would be lower by a large factor. Since the rate needs to be lower by 10^3, the electron distribution would be depressed by at least the same factor, relative to a Maxwellian distribution.

Although this analysis corresponds to an extreme case of a very high value of v_e^{inel}, even for the lowest value of v_e^{inel} of 10^5 s^{-1} the power input would be about 300 W, which is also rather high. Even in this case we would expect significant depression of the electron distribution.

Now we shall approach this problem from another direction. The rate at which electrons are fed in to the hole will be estimated directly. The electron distribution will be assumed to be "thermal," with a temperature of about $T_e \sim 2-5$ eV so that $v_e \sim 10^6$ ms^{-1}. In an ECR device the heating takes place in the center of the device, where all the electrons are energetically allowed to visit. The time between visits is, in the absence of collisions, $\tau_h \sim L/v_e$, where L, the system length is approximately 0.5 m and $\tau_h \lesssim 10^{-6}$ s. It is tempting to assume the electrons pick up $\Delta\varepsilon$ of several electron volts on each pass. Then for $n_e \sim 10^{18}$ m^{-3} the heating rate will again be too large. The rate is $(n_e/\tau_h)\Delta\varepsilon$ per unit volume, or up to about 10^6 Wm^{-3}. The reactor volume is less than 10^{-2} m^3, but this implies a power input of 10^4 W. However, if $\Delta\varepsilon \lesssim 1$ eV and can be positive or negative so that the heating involves a random walk in energy, then to gain about $E_{th}^{inel} \sim 5$ eV will take at least 25 passes through the heating zone. The heating rate is decreased by a factor of 25, to 400 W, which is high but not unreasonable. If the time between visits is determined by diffusion up and down the field line, the heating rate will be smaller still.

The implication of this analysis is that the maximum rate at which electrons are fed into the 'hole' cannot be with a frequency of more than $\nu^{ih} \sim P_{in}/(n_e\Delta\varepsilon V_r)$ (since $P_{in} \geq \nu^{ih} n_e \Delta\varepsilon V_r$, where $\Delta\varepsilon$ is about 10^{-18} J). This limits ν^{ih} to a few times 10^4 s^{-1}. The total rate is then $n_e\nu^{ih}$ or about 10^{22} m^{-3}s^{-1}. The density in the hole is this rate multiplied by the lifetime in the hole, which is $n_e\nu^{ih}/\nu_e^{inel}$ if inelastic collisions are responsible for removing the electrons. The ratio ν^{ih}/ν_e^{inel} could be from 0.1 to 10^{-4}, according to our estimates of ν_e^{inel}, which is the same factor found before by which the distribution is suppressed in the hole.

For the ICP a similar value of ν^{ih} is to be expected, but the heating process is somewhat different. The coolest electrons are not heated effectively, but the hot electrons may pick up an electron volt or more per bounce off the antenna-end of the reactor. Depending on the details, ν^{ih} may be even smaller and the hole may be filled even less effectively than in an ECR reactor.

4.11 Computation of the Electron Distribution

The electron distribution function is one of the most difficult aspects of the plasma behavior to calculate. For this reason we sometimes need to use a numerical approach, to obtain even a reasonable estimate of the distribution. In this section we use a simple numerical approach to finding the distribution. This method, as presented here, at first relies on averages over space but explicitly allows for the dependence of the distribution on $(v_z, v_\perp)$, the components of the velocity parallel and perpendicular to the z axis. It will also serve to introduce concepts that are needed in Chapter 8, where we show how to do a more general calculation, with space- as well as velocity-independent variables.

As a first example, suppose we are interested in the distribution function of electrons moving in a constant electric field E_z in the z direction. If there is no other electric field, this would correspond to a swarm experiment. In a positive column, there would be an additional electric field confining particles in the (radial) r direction, perpendicular to z. In terms of the simulation what this means is that in the positive column, particles at $r = 0$ with $|v_\perp|$ greater than some critical value have a chance of escaping to a wall. In the swarm we are not concerned with there being a wall.

The independent variables we shall use are $(v_z, v_\perp)$. To find the distribution function in terms of these we set up a "mesh" of discrete values: $(v_z = k\Delta v; v_\perp = j\Delta v)$ would be the simplest mesh we could use. The distribution $f(v_z, v_\perp)$ is then evaluated at mesh points (k, j). It may be helpful to think of the distribution as being a histogram, so that $f(v_z, v_\perp)$ is constant and equal to $f(k, j)$ in a cell from $v_z = (k - \frac{1}{2})\Delta v$ to $(k + \frac{1}{2})\Delta v$ and from $v_\perp = (j - \frac{1}{2})\Delta v$ to $(j + \frac{1}{2})\Delta v$.

4.11.1 Particle Motion

The calculation will be done by stepping in time until f settles down. See Fig. 4.1. It will be convenient to design the mesh and the time step Δt so that particles in the cell centered at $v_z = k\Delta v$ will move to the cell centered at $v_z = (k \pm 1)\Delta v$ in Δt. This is possible only because E_z is constant. In other words, the change in v_z is $\Delta v_z = \pm \Delta v$, the mesh size. Since

$$\Delta v_z = -\frac{e}{m} E_z \Delta t , \qquad (4.55)$$

we should probably choose Δv to be a convenient value, and then set $\Delta t = |\frac{\Delta v}{E_z}| \frac{m}{e}$, provided $v\Delta t \ll 1$ (see Chapter 8). Making Δv_z exactly match the mesh re-

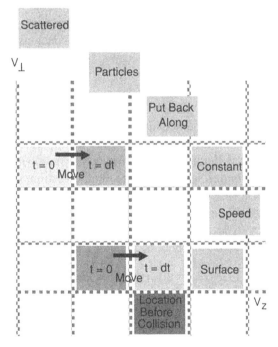

Fig. 4.1. A simple scheme for updating the distribution function. If E_z is constant, we can choose the time step Δt so that particles move exactly one cell in v_z in a time step Δt. $v_\perp$ does not change except during collisions, if we assume the electric field is only in the z direction. As a result, the density that was in velocity cell $v_z = k\Delta v$ will move to cell $v_z = (k+1)\Delta v$ after one time step. If the particles did not move in v_z by exactly Δv, that density would be shared between two v_z cells after Δt. Particles that collide go to a range of velocities. If the particles that collided end up having the same speed as each other, they will be put back in cells at the same speed v, where $v^2 = v_z^2 + v_\perp^2$. The more general case is described in Chapter 8.

duces "numerical diffusion," which is an artificial spreading out of the density that would occur if the particles from one cell had to be shared between a pair of cells.

4.11.2 Collisions

In each time step there is a chance of particles having a collision. Suppose the collisions are with a neutral background, and the collision frequency is $\nu \equiv \nu(\varepsilon)$, which depends on the electron's kinetic energy ε. Provided $\nu \Delta t$ is small, $\nu \Delta t \ll 1$, the fraction of the particles that collide in the time Δt is $\nu \Delta t$. This fraction of the particles in the cell must be removed after each time step and returned to other cells. Where the particles are put back after the collision depends on the type of collision. The description of collisions is discussed and illustrated in Chapter 8; for now we give a simple introduction.

Suppose that the kinetic energy of the particles is below the threshold for inelastic collisions, so that the collisions are all elastic. The elastic collision frequency is usually much bigger than all inelastic collision frequencies; thus most collisions are elastic even above threshold. The scattered particles must be returned at an appropriate energy and traveling in an appropriate direction.

For elastic collisions of electrons with neutrals, the electron's kinetic energy decreases very slightly. For now we neglect the small decrease in kinetic energy. Then the electron started with kinetic energy $\varepsilon_{\text{init}} = \frac{1}{2}m(\Delta v)^2[j^2 + k^2]$, and its final kinetic energy $\varepsilon_{\text{fin}} = \varepsilon_{\text{init}}$.

There are only a few cells on the mesh that exactly match this energy. For instance, for initial cell $(k, j) = (3, 4)$, particles could be put back at $(\pm 4, 3)$, as well as $(\pm 3, 4)$. $v_\perp$ is not allowed to be negative. There is also, for this particular set of integers, the pair of cells at $(\pm 5, 0)$ and the cell at $(0, 5)$, which exactly match the energy. This set of cells will not give an isotropic distribution of scattered particles (meaning that the distribution is not the same in all directions).

To see what an isotropic distribution should be, in the variables $(v_z, v_\perp)$, we need to know the expression for the element of area on the surface of a sphere. In spherical variables (r, θ, ϕ), or cylindrical variables (r, z, ϕ), the element of area (which in our case will be on a sphere of radius v) is

$$dS = r^2 \sin\theta \, d\theta \, d\phi = r \, dz \, d\phi. \tag{4.56}$$

The scattered particles are described by variables v_z, $v_\perp$, and ϕ_v, where ϕ_v is the azimuthal angle the velocity makes. They are on a surface of constant speed, $v = \sqrt{v_z^2 + v_\perp^2}$, and the element of area on this sphere dS_v is

$$dS_v = v \, dv_z \, d\phi_v. \tag{4.57}$$

Isotropy means that equal numbers of particles go through equal areas dS_v. In our computational problem, each cell corresponds to a range $\Delta\phi_v = 2\pi$; so at constant v, the number of particles in dS_v is proportional to dv_z. Since each cell has the same width Δv_z, we must put equal numbers of scattered particles in each v_z-cell, to achieve isotropy. We then choose the $v_\perp$-cells for those particles to conserve kinetic energy. See Fig. 4.1. This usually means that for each v_z we share the particles between a pair of adjacent $v_\perp$-cells, to get the correct average kinetic energy.

The set of variables $(v_z, v_\perp)$ was chosen to make the particle motion easier to handle accurately. The fact that particles are shared between cells with different $v_\perp$ when a collision occurs means that these variables are not optimal for handling collisions. The variables (v, μ), where $\mu = v_z/v$, are well-suited to handling collisions since an isotropic collision leaves particles at a speed v and equal numbers in equal ranges of μ. See Chapter 8 and Ref. [82].

Inelastic collisions only occur above a threshold energy, and they reduce the electron's kinetic energy. If the inelastic collision is an excitation, the decrease is by a fixed amount ε_{ex}. If the inelastic collision ionizes the atom, part of the first electron's energy is used to do the ionization and part is given to the newly released electron. The two electrons (which are indistinguishable) then each have a range of final energies. In all of these cases, we first decide how many electrons go to each final energy. Then, if the collision is isotropic (which is only likely to be approximately true), the electrons are returned to the mesh with equal numbers being put back in each Δv_z. Next, $v_\perp$ (or a pair of discrete $v_\perp$ values) is chosen for each v_z to achieve the correct kinetic energy.

4.11.3 Wall Losses

Since we are not keeping track of the particles' positions, we do not know when the particles are close to the wall, and so we cannot allow for wall losses exactly. Instead, we can calculate a probability that particles can hit a wall in a time step. Suppose the wall is at the end(s) in the z direction. Only particles with sufficient energy to hit the end can reach the wall. The distance d they travel in z in time Δt is $d = |v_z|\Delta t$. The probability of the energetic particles hitting the end will be estimated as d/L, where L is the length of the system. A similar estimate can be used if the losses are radial. The electrostatic potential is assumed to be a rectangular potential, for now.

4.11.4 Heating Mechanisms

In the case with the constant field E_z the heating mechanism is clear and the way it was included was already described. Two other cases will be considered here:

1. An ECR discharge, in which electron losses are to the ends (because of the magnetic field, which probably reduces radial loss of electrons to a very low level), uses ECR heating in the center of the discharge, where all the electrons can be heated. The heating gives a kick to $v_\perp$ each time the electron passes through the heating zone.
2. An inductive discharge has electron losses both to the ends and to the wall, and it uses heating at one end in z, where only hot electrons with (initially) high v_z can travel. Cold electrons in a plasma are often trapped by the electrostatic potential Φ. As they move around in the potential well, they slowly lose kinetic energy in elastic collisions. In a plasma in which the heating affects the hot

electrons but not the cold electrons, the cold trapped electrons would eventually cool down to the temperature of the neutral gas, T_g, unless there is some other process to heat them up. In the ICP, this process is Coulomb collisions.

All of these heating mechanisms can be included in a straightforward way. If the heating is only in a limited region, of length H, it is given a probability of happening equal to H/L.

4.12 The Energy Distribution of the Flux

This section turns to the ion distribution and gives examples of using a slightly different approach to kinetic calculations. The method relies on the idea that the ions are all falling in the same direction.

We will define the energy distribution function of the flux, the EDFF, $f_\Gamma(\varepsilon)$ so that the number of particles crossing a unit area per second, that are in the range of energies ε to $d\varepsilon$ is $f_\Gamma(\varepsilon)d\varepsilon$. Assume the particles are ions and that all the ions are moving in the positive z direction.

Inside a plasma with a constant electric field $\mathbf{E} = E_o\hat{\mathbf{z}}$, suppose first that singly charged ions are produced uniformly in space, with zero initial energy. Collisions are neglected for now.

We shall now calculate $f_\Gamma(\varepsilon)$, given that the ionization rate is S_o per unit volume per second. To do this, first notice that particles that have fallen a distance z have kinetic energy $\varepsilon = eE_oz$. The number of particles (per unit area, per second) produced between z and $z + dz$ is S_odz. These particles, in one second, with energies between $eE_o(z_f - z)$ and $eE_o(z_f - z - dz)$, pass through a point at a position z_f "downstream," in steady state. Since the range of energies of these particles is $d\varepsilon = eE_odz$, the number passing per second in the range $d\varepsilon$ is $S_odz = S_o\frac{d\varepsilon}{eE_o}$. This means that $f_\Gamma(\varepsilon) = S_o/eE_o$, at least up to some maximum kinetic energy $\varepsilon_{\max}$, which is set by the length of the plasma.

The relation between $f_\Gamma(\varepsilon)$ and the usual distribution $f(\varepsilon)$ is that if the motion is in the z direction with speed v_z, $f_\Gamma(\varepsilon) = v_z f(\varepsilon)$. (The flux is equal to the density multiplied by the velocity.) In this case, therefore, $f(\varepsilon) = S_o/eE_ov_z$, or

$$f(\varepsilon) = \frac{S_o}{eE_o\sqrt{2\varepsilon/m}}. \tag{4.58}$$

Now suppose the particles are all produced at $z = 0$, but that they undergo collisions with a mean free-path λ, and that after the collisions their kinetic energy is reduced to zero. We shall see that this is actually somewhat similar to the previous calculation. All the particles launched at $z = 0$ have to pass through any other point $z > 0$. Hence the number (per unit area) passing a point per second is the number launched per second, per unit area, at $z = 0$, which will be denoted S_o. The number of collisions per unit length is thus S_o/λ. This allows a simple calculation to be done. First, to show how an alternative approach might work, we can calculate the EDFF, iteratively, in a region of width L, where $\lambda \lesssim L$, as follows. (The analysis up to and including Eq. (4.62) can be omitted.)

The lowest-order EDFF consists of the particles that have had no collisions. These particles all have energy $\varepsilon = eE_oz$. The number that have not collided decays like $e^{-z/\lambda}$:

$$f_\Gamma^{(o)} = S_o e^{-z/\lambda} \delta(\varepsilon - eE_oz) \qquad (4.59)$$

since we defined S_o to be the number per unit area, per second, that are launched.

The particles that have collided once provide a secondary "source rate":

$$S_1 = \frac{S_o}{\lambda} e^{-z/\lambda}. \qquad (4.60)$$

This is the number of particles from the initial beam of particles that collide, per unit volume per second. The source rate of particles having their second collision is S_2, and so on. Note that S_o is a rate per unit area per second, whereas S_1, S_2, etc. are per unit volume per second.

At any point z, the source at some other point, z', $S_1(z')$, contributes particles that have fallen, from z' to z. Their energy is $\varepsilon = eE_o(z - z')$.

We need the flux of particles at point z, that have collided once, with kinetic energy between $\varepsilon = eE_o(z - z')$ and $\varepsilon + d\varepsilon = \varepsilon E_o(z - z' - dz')$. The number produced per second per unit length near z', after one collision, is $\frac{S_o}{\lambda} e^{-z'/\lambda}$.

These particles come from near z', in a range of distances $(-dz')$. The fraction of these particles that make it to point z without another collision is $e^{-(z-z')/\lambda}$, and so the flux passing z in this energy range is $\frac{S_o}{\lambda}(-dz')e^{-z/\lambda}$. Now $-dz' = d\varepsilon/eE_o$; so this part of the flux, in the energy range $d\varepsilon$ at ε, is $f_\Gamma^{(1)} d\varepsilon$, where $f_\Gamma^{(1)} = \frac{S_o}{\lambda eE_o} e^{-z/\lambda}$, for $\varepsilon < \varepsilon_{\max} = eE_oz$.

The source rate due to this is (minus) the flux of these particles differentiated with respect to z. The flux of these particles is the energy integral of the flux per unit energy, $f_\Gamma^{(1)}$. ($f_\Gamma^{(1)}$ is the EDFF of particles that have collided once; $f_\Gamma^{(n)}$ is the EDFF of particles that have collided n times.) The energy integral is just $\varepsilon_{\max} = eE_oz$ times $f_\Gamma^{(1)}$. The source rate due to particles having a second collision is then

$$S_2 = \frac{\varepsilon_{\max} f_\Gamma^{(1)}}{\lambda} = \frac{zS_o}{\lambda^2} e^{-z/\lambda}. \qquad (4.61)$$

Now the total source rate is just $\frac{S_o}{\lambda}$, since the total flux passing a point is S_o, in steady state. It is likely, therefore, that $S_3 = \frac{Sa_o}{2\lambda^3} z^2 e^{-z/\lambda}$, and so on, so that

$$S = S_1 + S_2 + S_3 + \cdots = \frac{S_o}{\lambda}\left[1 + \frac{z}{\lambda} + \frac{z^2}{2\lambda^2} + \cdots\right]e^{-z/\lambda}$$

$$= \frac{S_o}{\lambda}[e^{z/\lambda}]e^{-z/\lambda} = \frac{S_o}{\lambda}. \qquad (4.62)$$

We will now return to the "simple" approach, and say that the total source due to particles that immediately beforehand had a collision is S_o/λ, per unit volume per second. At any point (ε, z) the flux between ε and $\varepsilon + d\varepsilon$ is the flux coming from between a distance $z_1 = \frac{\varepsilon}{eE_o}$ and a distance $z_2 = \frac{\varepsilon + d\varepsilon}{eE_o}$, upstream. That flux is

$$\frac{S_o}{\lambda}(z_2 - z_1)e^{-z_1/\lambda} = \frac{S_o d\varepsilon}{\lambda e E_o}e^{-(\varepsilon/eE_o)/\lambda}, \qquad (4.63)$$

or

$$f_\Gamma(\varepsilon) = \frac{S_o}{\lambda e E_o}e^{-\varepsilon/eE_o\lambda}; \varepsilon < \varepsilon_{\max}. \qquad (4.64)$$

This neglects the source, S_o, per unit area per second, at $z = 0$. As noted above, this source contributes $f_\Gamma^{(o)}(\varepsilon) = S_o e^{-\varepsilon/eE_o\lambda}\delta(\varepsilon - eE_o z)$. The total energy distribution function of the flux is thus

$$f_\Gamma(\varepsilon) = S_o e^{-\varepsilon/eE_o\lambda}\left(\frac{1}{\lambda e E_o} + \delta(\varepsilon - eE_o z)\right); \varepsilon \le \varepsilon_{\max}. \qquad (4.65)$$

The integral of this over energy should equal S_o:

$$\Gamma = \int_0^{\varepsilon_{\max}} f_\Gamma d\varepsilon = S_o \int_0^{\varepsilon_{\max}} e^{-\varepsilon/eE_o\lambda}\left(\frac{1}{\lambda e E_o} + \delta(\varepsilon - eE_o z)\right) d\varepsilon. \qquad (4.66)$$

This in turn gives

$$\Gamma = S_o\left([-e^{-\varepsilon/eE_o\lambda}]_0^{\varepsilon_{\max}} + e^{-z/\lambda}\right) = S_o. \qquad (4.67)$$

Far enough downstream, the initial beam is obliterated, and only the first term remains. The upper limit is irrelevant far downstream because $\varepsilon_{\max} \gg eE_o\lambda$.

Exercises

1. For the distribution shown in Fig. 4.2, plot the mean flow velocity versus z, if the origin of coordinates in phase space, $z = 0$, $v_z = 0$, is at point A. Repeat, assuming the origin is at point B, and finally if the origin is at point C.

2. In an ECR etcher, the magnetic field in the resonance zone is usually 875 gauss, or 8.75×10^{-2} T. Suppose the electron temperature, T_e, is 5 eV and that the electrostatic potential in volts in the main, quasineutral, plasma varies with radius r as

$$\Phi = -T_e\left(\frac{r}{a}\right)^2, \qquad (4.68)$$

where $a = 0.05$ m is the reactor radius. Calculate the $\mathbf{E} \times \mathbf{B}$ drift velocity, $\mathbf{v_E} = \frac{\mathbf{E} \times \mathbf{B}}{\mathbf{B}^2}$.

If the ion density in the sheath is 10^{11} cm^{-3}, estimate the sheath thickness x_s needed for a sheath drop of $4T_e$. Assuming that the sheath electric field is about $\frac{4T_e}{x_s}$, calculate $\mathbf{v_E}$ in the sheath. Compare the ion and electron gyroradii

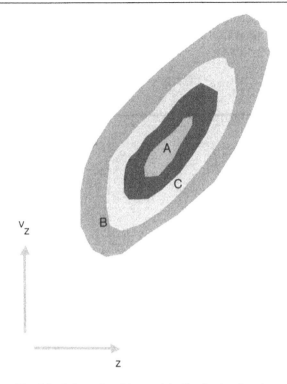

Fig. 4.2. Schematic of the particle distribution function.

to the sheath thickness, x_s. Assume the ions are He^+, and make reasonable assumptions about $v_{\perp i}$ and $v_{\perp e}$.

3. Suppose that a mechanism is available, to heat electrons in a plasma, provided the electrons are in a narrow range of kinetic energies, ε_h to $\varepsilon_h + \Delta\varepsilon$. The heating rate is constant, within this range, and zero otherwise. If the threshold for ionization is $\varepsilon_{th} = 15$ eV, and $\Delta\varepsilon = 1$ eV, how will the ionization rate depend on the value of ε_h? The electron elastic mean free path is $\lambda_e^{el} = 10$ cm. The mean free path for ionization, for electrons above threshold, is $\lambda_e^{ion} = 100$ cm. The plasma is contained in a cylinder of radius $a = 20$ cm and height $h = 40$ cm. Describe the time evolution of the electron distribution and the electrostatic potential, if the heating is very fast in the range from ε_h to $\varepsilon_h + \Delta\varepsilon$.

4. Calculate the density of ions, n, in terms of distance w from a boundary, if the electrostatic potential is

$$\Phi = -\Phi_0 (1 - \omega/\omega_m)^2, \qquad (4.69)$$

the ion mobility is μ, the ionization rate S is

$$S = S_0 (1 - (1 - \omega/\omega_m)^2), \qquad (4.70)$$

and the diffusion is one dimensional, with w being a Cartesian coordinate. Here ω_m is the distance from the boundary at which the density has its maximum. Neglect diffusion, so that the flux per unit area of ions is $\Gamma = n\mu E$, where E is the electric field.

Find n, by first setting $x \equiv \omega/\omega_m$ and showing that the electric field, in the direction toward the wall, is

$$E = \frac{d\Phi}{d\omega} = \frac{2\Phi_0}{\omega_m}(1-x).$$ (4.71)

Then show that the flux per unit area is

$$\Gamma = \int_\omega^{\omega_m} S \, d\omega = \omega_m S_0 \int_x^1 (2x - x^2) dx$$ (4.72)

$$= \frac{\omega_m S_0}{3}[2 - 3x^2 + x^3].$$

Finally, show that

$$n = \frac{\omega_m^2 S_0}{6\mu\Phi_0}\left(\frac{2 - 3x^2 + x^3}{1-x}\right).$$ (4.73)

What is n at $x = 0$ and at $x = 1$? To find $n(x = 1)$, set $y = 1 - x$ and take the limit as $y \to 0$.

This problem is intended to illustrate how n is largest where Γ goes to zero. Comment on the assumptions about Φ and S. Does the solution for n seem to be consistent with Φ? Refer to the derivation of the ambipolar diffusion coefficient and to the expected relation between n and Φ.

5. An electron gas is contained in an infinite cylinder of radius a. The electrostatic potential is zero in the interior and drops to a negative value, $-\Phi_w$, at the wall. If the electrons have a density n_e, which we shall assume is uniform in space, and are at a temperature T_e, and they have an isotropic distribution, what is the flux of electrons per unit of the wall?

    The cross section for electron–neutral collisions that lead to ionization of the neutral is $\sigma = \sigma_0 u(\varepsilon - \varepsilon_{th})$. Here ε_{th} is the threshold kinetic energy for ionization and u is a step function; thus $u(\varepsilon > \varepsilon_{th}) = 1$ and $u(\varepsilon < \varepsilon_{th}) = 0$. What is the ionization rate per unit volume?

    In steady state, the electron loss rate per unit length should be equal to the ionization rate per unit length. What must the relation be, between T_e and ε_{th}, for these rates to be equal?

5

Plasma Chemistry

5.1 Plasma Chemistry

The chemical processes that can take place in a plasma are exceptionally complicated and little understood. A treatment of all the possible chemistries of interest for semiconductor fabrication would be an enormous undertaking. Here we establish some guidelines for thinking about plasma chemistry, rather than attempting to describe all the possibilities. A number of texts on plasma chemistry are available, although the emphasis in many is more on plasma polymerization than on silicon processing [83–91].

The term plasma chemistry is usually not appropriate to describe the important effects. The hot electrons from the plasma are responsible for driving unusual chemistry – mostly neutral chemistry. (The main exception to the electrons driving the chemistry is activation of surfaces by ions.) Perhaps only the first step in a chain of reactions even involves the electrons, however. The energy an electron needs to ionize a neutral is much higher than the energy needed to dissociate most neutrals. There are usually vastly more electrons with enough energy to dissociate neutrals than there are electrons that have enough energy to ionize neutrals.

The typical dissociation energy might be 5 eV, whereas the typical ionization energy is probably at least three times this. If the electrons themselves are very hot, with a temperature T_e corresponding to about 5 eV, then the expected number at 5 eV will be about $n_o e^{-\varepsilon/T_e} \simeq n_o e^{-1}$. The number at 15 eV will be about $n_o e^{-3}$. Since the ratio of these numbers is only an order of magnitude, in such a hot plasma the ratio of the cross sections will determine whether ionization is close to being as frequent as dissociation. Suppose instead now that $T_e \simeq 2$ eV. Then the number of electrons at 5 eV is now about $n_o e^{-2.5}$, whereas the number at 15 eV is about $n_o e^{-7.5}$. The number of electrons that can ionize is lower by e^{-5}, which is about 3×10^{-3}. Even if the ionization cross section is large, most of the processes driven by the electrons are dissociations in this case. This argument shows that if $T_e \gtrsim 5$ eV (which is unusual; it can happen in ECR systems) ion chemistry becomes more important. Usually, electron-driven neutral chemical reactions are most frequent. After the electron temperature, the next most important factor in determining the chemical reactions taking place is probably the state of the wall, including the wall temperature.

An accurate quantitative description of a processing plasma is made difficult to obtain by the absence of much of the important information about chemical reactions, both in the gas phase and on the surface. Building up a picture of the important chemical processes taking place during plasma etching or deposition typically depends on testing simple models against experimental data. Sophisticated numerical models can also be used to predict the sensitivity of a process to the reaction rates, for instance. Once the critical reaction steps have been tentatively identified it may be possible to infer the rate constants of the dominant process, by comparing experimental measurements to the corresponding computed values. In this discussion of plasma chemistry the emphasis will be on the use of straightforward reasoning based in part on experiments to explain and predict the chemical densities and fluxes to surfaces.

In this chapter, rates and types of reaction are discussed. Next, several of the more commonly used etching and deposition chemistries will be described briefly, in part to serve as worked examples. The interaction of the chemistry with the type of reactor can be significant. The main reactors used at present are ICPs (meaning Inductively Coupled Plasmas), ECR reactors, (ECR meaning Electron Cyclotron Resonance) and capacitively coupled reactors. The ICP and ECR devices tend to have high densities of charged particles, $n \lesssim 10^{12}$ cm^{-3}, and most of the electrons in these devices can be described as having a relatively high temperature T_e of 2–5 eV. In capacitively coupled reactors the density is usually lower and T_e may be lower.

We shall consider in this chapter and the next an example of CF$_4$ being used as the feed gas in an ECR etcher. CF$_4$ is an exceptional etching gas for which more of the relevant information about gas phase chemistry is available than for most other etching gases. What takes place on the surface will depend on the nature of the surface but is much less well known even than the gas-phase processes. The fact that the ECR system can sustain a high density of hot electrons is critical to the performance of the reactor because of the effect of these many hot electrons on the negative ion population. Hot electrons are less likely than cold electrons to combine with (attach to) neutrals to form negative ions, and they are more likely to knock electrons off negative ions. The high number density of hot electrons makes the CF$_4$ in the ECR system electropositive (with mostly positive ions), whereas with fewer hot electrons the CF$_4$ is electronegative (having many negative ions and relatively few electrons). This pattern can also be expected to hold for other gases that tend to form negative ions.

After noticing that the hot electrons suppress formation of negative ions and that the gas phase chemistry is better known than the surface chemistry, we must collect more information about the chemical reactions that can take place and about the results of relevant experiments. The known gas phase chemical reactions relevant to a CF$_4$ plasma are summarized in Section 5.6, in part to illustrate the types of processes which must be considered.

5.2 Rates of Reaction

We will divide the different types of reaction into two classes depending on how the rate of the reaction is specified. In the first case the reaction rate, which is the number of reactions per unit volume per second, is equal to a reaction rate coefficient r (which depends on a temperature, $r \equiv r(T)$) multiplied by the number densities of the reacting species. In the second case the reaction rate is specified in terms of a cross section and the cross section typically depends in a complex way on the energy of the individual reacting particles.

The distribution of kinetic energies that the electrons in a discharge have is often not described accurately by saying they have a temperature T_e equal to some value. If they were at a temperature T_e their distribution with respect to energy would have a Maxwellian form, and the number of particles per unit kinetic energy ϵ would be

$$f = A\epsilon^{1/2} \exp(-\epsilon/k_B T_e), \tag{5.1}$$

where k_B is Boltzmann's constant and $\epsilon \equiv \frac{1}{2}m_e v_e^2$, where v_e is the electron speed and m_e is the electron mass. The neutral particles, in contrast, will typically have kinetic energies that do have a nearly Maxwellian distribution and so the neutrals have a temperature T. If we know the distribution corresponds to some temperature we can average over all the particle energies to find an overall reaction rate coefficient $r(T)$ that depends on temperature rather than on individual particle energies. For the electrons in particular this averaging may not be possible since the number of electrons in each range of energies may not be known.

For two neutral species A and B with numbers per unit volume of n_A and n_B respectively, the rate per unit volume of a reaction between them, R_{AB}, will usually be proportional to both n_A and n_B:

$$R_{AB} = r_{AB}(T)n_A n_B. \tag{5.2}$$

(This expression is appropriate if one molecule of A reacts with one molecule of B. If, for instance, two molecules of A reacted with one molecule of B, then R_{AB} would be proportional to $n_A^2 n_B$.) This overall rate is an average over the energies of the particles, all of which are assumed to have temperature T.

Reactions with more than two species are possible, but the discharges we are interested in are at low pressures and the low number densities of particles in the gas phase make it unlikely that three or more molecules will all be at the same place at the same time to be able to react. In a reaction between species A, B, and C, called a three-body reaction, the rate would be

$$R_{ABC} = r_{ABC}(T)n_A n_B n_C. \tag{5.3}$$

r_{ABC} is typically so small that the densities n found in low-pressure discharges do not compensate for this and the rate R_{ABC} is also very small. "Three-body" reactions may be important on the walls where particles can remain in place long enough for other particles to have a better chance of reacting with them.

For reactions involving electrons or ions, we typically have to deal with a cross section σ for the reaction, and σ usually depends strongly on kinetic energy. As described earlier, any particle of species A that falls within the area σ_{eA} (which is the cross section for electron collisions with species A) around an electron is supposed to collide with it; so when the electron moves a distance d, it sweeps out a volume $\sigma_{eA}d$ and any particle of species A, the center of which is in this volume, undergoes a collision with the electron. The density of species A is n_A and hence the volume that contains one particle of species A is $1/n_A$ on average, and this is the average volume the electron must sweep out to have a collision. The volume swept out in going a distance d is $\sigma_{eA}d$. Therefore, if the average distance between collisions is λ then the average volume swept out between collisions by an electron is

$$\sigma_{eA}\lambda_e = n_A^{-1}. \qquad (5.4)$$

This means that $\lambda_e = (n_A\sigma_{eA})^{-1}$. (It is a common mistake to use the wrong density here; λ_e is the electron mean free path but n_A is the density of species A.)

The electron is moving at a speed v_e with a kinetic energy $\epsilon = \frac{1}{2}m_e v_e^2$. The average time between collisions must be $\tau_c \equiv \lambda_e/v_e$ and the collision frequency is

$$\nu_{eA} \equiv 1/\tau_c = n_A\sigma_{eA}v_e. \qquad (5.5)$$

This is the frequency at which a single electron can be expected to collide with (and perhaps ionize or excite, depending on the type of collision) species A. The reason that the speed of the particles of species A was not included is that they are typically so slow moving compared to the electrons that they are effectively stationary from the point of view of the electron. If this is not true the formula must be corrected.

To find the total rate of electron reactions we have to sum all the rates of reaction of all the electrons. The electron energy distribution function $f(\epsilon)$ is the number of electrons per unit volume and per unit kinetic energy, evaluated at the kinetic energy ϵ. The number of electrons per unit volume in the range of kinetic energy $d\epsilon$ at kinetic energy ϵ (that is, between ϵ and $\epsilon + d\epsilon$) is $f(\epsilon)d\epsilon$. If each electron collides with species A at a rate $\nu_{eA}(\epsilon)$ then the total rate of collision with species A (per unit volume) for the electrons in the range $d\epsilon$ at ϵ is $\nu_{eA}(\epsilon)f(\epsilon)d\epsilon$. To sum over all the electrons we must integrate this over all kinetic energies to get the total rate of reaction of electrons with species A, denoted R_{eA}:

$$R_{eA} = \int_0^\infty \nu_{eA}f(\epsilon)d\epsilon = n_A \int_0^\infty \sigma_{eA}(\epsilon)v_e(\epsilon)f(\epsilon)d\epsilon. \qquad (5.6)$$

We shall now examine this collision process from the point of view of species A. The total reaction rate (number of reactions per unit volume per second) between electrons and species A can be written

$$R_{eA} = n_A\nu_{Ae}. \qquad (5.7)$$

The frequency ν_{Ae} is the frequency with which a particle of species A reacts with an electron and is not likely to equal ν_{eA}. The study of chemistry in the discharge involves estimating ν_{Ae}, that is,

$$\nu_{Ae} = \int_0^\infty \sigma_{eA}v_e(\epsilon)f(\epsilon)d\epsilon \qquad (5.8)$$

for many different cross sections. To do this we need to know $f(\epsilon)$ for the electrons.

In the ECR discharges used in plasma etching a frequent observation (obtained primarily from probe measurements) is that the electron distribution function consists approximately of two Maxwellian distributions:

$$\begin{aligned} f(\epsilon) &= A_1 F_M(T_1) + A_2 F_M(T_2) \\ &= A_1\epsilon^{1/2}\exp(-\epsilon/k_B T_1) + A_2\epsilon^{1/2}\exp(-\epsilon/k_B T_2). \end{aligned} \qquad (5.9)$$

The number of electrons with each temperature T_1 and T_2 is controlled by the constants A_1 and A_2. The total frequency ν_{Ae} is therefore

$$\nu_{Ae} = A_1 \int_o^\infty \sigma_{eA} \nu_e \epsilon^{1/2} e^{-\epsilon/k_B T_1} d\epsilon + A_2 \int_o^\infty \sigma_{eA} \nu_e \epsilon^{1/2} e^{-\epsilon/k_B T_2} d\epsilon, \quad (5.10)$$

and similarly for more than two Maxwellians. The integrals over energy can be evaluated for many temperatures and then stored for future use. The results of doing this are given in Ref. [92] for the main electron–neutral reactions in a CF_4 plasma.

We have outlined the two main ways in which we calculate reaction rates. Next we turn to the main kinds of reaction that take place, starting with the gas phase.

5.3 Types of Reaction

In addition to the types of collision that occur in a simple gas, in a gas consisting of more complex molecules the molecular collisions include more complex varieties of i) electron–molecule collisions and ii) heavy-particle collisions:

i)	Ionization:	$e + XY \rightarrow 2e + XY^+$.
	Attachment:	$e + XY \rightarrow XY^-$.
	Dissociation:	$e + XY \rightarrow e + X + Y$.
	Polar Dissociation:	$e + XY \rightarrow e + X^+ + Y^-$.
	Dissociative	
	– Ionization:	$e + XY \rightarrow 2e + X + Y^+$.
	– Recombination:	$e + XY^+ \rightarrow X + Y^*$.
	– Attachment:	$e + XY \rightarrow X + Y^-$.
	Electron Impact Detachment:	$e + XY^- \rightarrow 2e + XY$.
ii)	Charge Transfer	
	– Resonant:	$X^+ + X \rightarrow X + X^+$.
	– Non resonant:	$X^+ + Y \rightarrow X + Y^+$.
	Recombination:	$X^+ + Y^- \rightarrow X + Y$.
	Associative Detachment:	$X + Y^- \rightarrow XY + e$.
	Excitation Transfer:	$X + Y^* \rightarrow X^+ + Y + e$.
		$\rightarrow XY^+ + e$.
		$\rightarrow X^* + Y$.

From the point of view of CF_4 being pumped into a chamber containing hot electrons, the first important reactions that are likely to occur involve collisions with electrons. The possible types of collision include ionization of the CF_4 by loss of an electron, attachment of electrons to the CF_4, and break up of the CF_4 into other neutral or charged species or combinations of neutral and charged species. These are all listed in Refs. [92]–[94]. (The computation of cross sections for electron collisions with neutrals is discussed in Ref. [95].)

The next important category of gas phase reactions could be taken to be neutral–neutral reactions, listed in Refs. [92]–[94]. There are fewer important neutral–neutral reactions in this system. These neutral reactions are assumed to be between neutral particles at the same temperature.

Finally, reactions of ions with neutrals must be included. These reactions of ions with neutrals involve ions that typically have fallen some distance in a strong electric field so the ion may be moving fast. A "charge-exchange" reaction of a positive ion for instance may allow the ion to simply pull an electron off a neutral with little other effect. The fast-moving ion then becomes a fast-moving or "hot" neutral and the slow neutral, which lost the electron, becomes a slow ion. The charge exchange reactions of ions with neutrals are probably the main cause of ions changing their species. They also produce enough hot neutrals to affect processes on the surface.

Elastic collisions of ions with neutrals are also possible, as are elastic collisions of electrons with neutrals and of neutrals with each other. When electrons have elastic collisions with neutrals the electrons transfer a very small fraction of their kinetic energy to the neutral. However, because the electrons also suffer a large change in momentum elastic collisions with neutrals have a strong tendency to keep the electron distribution isotropic (meaning that the motion is isotropic: isotropic motion is no more likely to be in one direction than another.) Elastic neutral–neutral collisions are the most common collisions neutrals undergo and they tend to maintain the neutrals' isotropy and their Maxwellian distribution. Positive ions do not have an isotropic motion since they tend to fall out of the plasma as the electric field pulls them to the walls. Both charge-exchange collisions and elastic collisions of ions are with neutrals. Elastic ion–neutral collisions tend to randomize the ion motion. Charge-exchange collisions reduce the ion velocity to nearly zero, and so if the ions are falling in a field with a constant direction, charge-exchange collisions reduce the ions' sideways velocity. However, plasma etchers are kept at low pressure largely to prevent the randomization of the ion velocity; thus we can expect the ions to move primarily outward toward the wall.

5.4 Etching Recipes

The processes that are used at low gas pressure to remove material from a surface are:

1. Sputtering. (This produces about one atom per ion, which is a low yield. It is not selective, and it can remove nonvolatile products.)
2. Chemical etching. (This is selective but etches isotropically.)
3. Chemical etching where ions provide energy to activate the chemical reaction; or where ions remove an inhibitor, where the inhibitor prevents etching.

Inhibitors

Inhibitors are usually carbon halides, which form Teflon-like coatings. The type of etch depends on the relative size of the fluxes of etchant Γ_{etch} and inhibitor, Γ_{inhib}:

$\Gamma_{etch} \gg \Gamma_{inhib}$ gives isotropic etching.

$\Gamma_{etch} \ll \Gamma_{inhib}$ gives film deposition.

$\Gamma_{etch} \sim \Gamma_{inhib}$ gives the possibility of anisotropic etching.

(O_2 is added to react with inhibitors.)

Silicon Etching by Halogens

F_2 etches Si but gives a rough surface. Fluorine plasma etching relies on ions to supply energy, which can boost the etch rate by an order of magnitude over the rate of purely chemical etching.

CF_4 tends to break up in a plasma because all its excited electronic states are unstable, which includes CF_4^+. (Its vibrational states have energies of around $1/10$ eV.)

Carbon with a moderate amount of fluorine forms a film, whereas an abundance of fluorine with little carbon etches the silicon. As before, the inhibitor to etchant flux ratio is crucial. Having but a little inhibitor means isotropic etching, which happens when there is more than three times as much F as C. However, too much inhibitor gives a film; if there is less than twice as much F as C, all the surfaces are coated and no etching occurs. But when there is between two and three times as much F as C, the ions can clean the trench bottom but not the sidewalls and very anisotropic etching can occur.

Adding oxygen has different effects depending on the amount of oxygen. A little O_2 burns the inhibitor, as mentioned above. More O_2 can oxidize the silicon surface, slowing etching. A lot of O_2 simply dilutes the fluorine.

Adding hydrogen removes F, pushing the equilibrium toward film formation. (Chlorine etches silicon, but the etching depends on the crystal plane and the doping. Cl_2 chemisorbs, giving ineffective etching. Cl forms a layer of $SiCl_x$ with a low etch rate unless the doping is n^+, which means heavily n-type. Ion-assisted etching by either chlorine species is possible.)

SiO_2 Etching by CF_x

The etching is driven by ion energy (so it does not depend on the substrate temperature) and is anisotropic. An ion energy of 500 eV can give an etch rate of a few thousand Angstroms per minute. Since F etches Si and SiO_2 it gives no selectivity; consequently, CF_x is used [28]. CF_x with $x < 4$ or CF_4/H_2 have a low enough F density to give selective etching. The effects of adding H_2 and O_2 are as usual. Adding O_2 gives more F, causing the etch rate to go up but eliminating selectivity.

A combined flux of ions and CF_x forms a layer containing C; the C can etch SiO_2. The O in the SiO_2 prevents film buildup (by "burning" the film) so that SiO_2 etches but not Si. (With too much H_2, thick films grow even on SiO_2.)

Photoresist

Photoresist is usually an organic polymer, made of C and H. O_2 plasma etches photoresist isotropically. O_2 chemical etching is selective over Si and SiO_2. C_2F_6/CF_4 speeds the chemical etch.

Al Etching by Halides

F is not useful because AlF_3 is involatile. Cl_2 (and Cl to a lesser extent) etches Al; CCl_4, $SiCl_4$, etc. are added as inhibitors, which makes ion-enhanced etching anisotropic. Al_2O_3 is not etched by Cl_2 or Cl; physical sputtering or the inhibitors CCl_4 or $SiCl_4$ may be used. (Water vapor has to be scavenged, by BCl_4 or $SiCl_4$.)

Si_3N_4 as a Mask/Dielectric

Chemical F etching is isotropic and selective over SiO_2 but not Si. CF_x etching driven by ion energy is selective over Si and photoresist but not SiO_2.

5.4.1 Current Understanding of Etching Processes

Perhaps the most revealing studies of etching mechanisms have been done using ion and molecular beams, starting with the work of Coburn and Winters [96, 97]. This appears to have prompted a large number of studies [98–106]. For a discussion of recent work on Cl etching of polysilicon, see Ref. [107], where it was confirmed that the etching yield shows a strong dependence on the ratio of the neutral to ion fluxes. Sticking probabilities of Cl_2 on Si [108] and other materials [109] were also obtained. Fluorocarbon etching of Si was studied in Ref. [110]. Reincidence of byproducts in metal etching was stressed in Ref. [111]. Recent work on fluorocarbon plasma etching of Si and SiO_2 has been described in Ref. [28 and 112].

5.5 The Chemistry of Film Deposition Using Plasmas

Plasmas can deposit films on surfaces that are relatively cool, which means the surface is less likely to be damaged; and some of the films made using plasmas cannot be deposited otherwise. Low-temperature PECVD films tend to be amorphous, whereas Chemical Vapor Deposition (CVD) can make crystalline films.

CVD is the deposition of a film on a surface by chemical reactions in a gas and on the surface, where the reactants are all at similar or the same temperature(s) and that temperature is high enough to drive the chemical reactions.

PECVD (Plasma Enhanced Chemical Vapor Deposition) is the deposition of a film on a surface by reactions in a gas and on the surface, where the reactions are driven to a significant extent by the energy of the charged (plasma) particles. Usually the electrons' kinetic energy is used to dissociate the input gas.

PECVD requires at least 0.1 to 10 Torr to allow a chance for gas phase reactions. Thus the mean free path for neutrals is $\lambda \lesssim 1$ mm. The surface temperature is not used to activate reactions so it is not a crucial factor in the rate, but it does affect film properties. The high rates that occur mean films deposit right away – perhaps near the inlet – thereby making uniformity a problem.

5.5.1 PECVD Recipes

Three types of deposited film containing silicon are of considerable importance for microelectronics: a–Si:H (amorphous silicon), SiO_2 (silicon dioxide, which is a glass), and silicon nitride.

a–Si : H

Amorphous silicon is often deposited using silane, SiH_4. SiH_4 explodes in air or water vapor. In discharges it forms SiH_3^+, SiH_3^-, and SiH_2^- (so the discharge can be electronegative). All the (—H) bonds have energy of about 3 eV. Kushner [113, 114] lists a) a large number of electron impact reactions, b) neutral–neutral reactions, c) I^+–n (positive ion–neutral) reactions, d) e–I^+ (electron–positive ion) reactions, and e) I^+–I^- (positive ion–negative ion) recombinations.

Kushner's model of the surface indicates that SiH_2 attaches to anywhere on a silicon surface, but the process makes a rough surface. SiH_3 only attaches to "active sites" with "dangling bonds," which the ion bombardment creates. SiH_3 diffuses along the surface to these active sites, filling in the roughness. To get a good film we need active sites (hence ion bombardment) and an abundance of SiH_3 compared to SiH_2 to fill in the roughness, and we need the SiH_3 to be free to move (having a high surface diffusion coefficient) to get to the active sites. (Surface reaction probabilities in a–Si:H have been found from decay rates in an "afterglow" [115].)

SiO_2

The common ways to make SiO_2 are:

1. Oxidation of Si at about 1,000°C in O_2 or H_2O.
2. CVD at about 700°C using an SiH_4/O_2 or TEOS/O_2 mixture. PECVD, with the same gases, requires 100–300°C.

SiH_4 is used with Ar as a carrier and N_2O, NO, or O_2 to provide oxygen, generating SiH_3, SiH_2, and O radicals. SiH_3 sticks to O on the surface, and the O removes H. Because the sticking coefficient is high, the film is not uniform.

TEOS, $Si(OC_2H_5)_4$, is an inert liquid. (The (—O) bonds have energy of about 4 eV.) Inert gases (N_2 or Ar) are used to carry the vapor consisting of O_2 (which removes C and H from TEOS) with about a percent of TEOS. Because the sticking coefficient is low, films may be uniform.

Silicon Nitride

Silicon nitride is usually made from SiH_4/NH_4 at about 1 Torr and 400°C. Electron impact creates SiH_3, SiH_2, and NH. Heat or ion bombardment activate desorption of H. NH_3 gives the best films, but nevertheless with a lot of H in the film.

5.5.2 Reactive Sputtering

In reactive sputtering, a feed gas is dissociated by electrons and reacts with the sputter target and the substrate, in addition to ions sputtering material off the target. An example is an Si target, with O_2 gas, which can give a better deposited SiO_2 film composition than an SiO_2 target by itself.

Reactive sputtering is used to deposit ceramics (such as oxides and nitrides), using O_2 or H_2O to provide O; CH_4 or C_2H_2 to provide C; N_2 or NH_3 to provide N; and SiH_4 to provide Si. If the ion flux is relatively high, the target probably stays uncovered ("metal" mode). If the neutral gas flux is relatively high the target can be covered ("covered" mode), which yields a lower deposition rate.

5.6 The Gas Mixture in a Plasma: The Case of CF$_4$

To illustrate the issues arising in studying plasma chemistry, it is useful to focus on a specific system. We will now attempt to use CF$_4$ data and some experimental results to estimate the composition of the CF$_4$ plasma and neutral mixture. First we summarize some important results to set the scene for what follows.

Data for CF$_4$

Sources for the relevant cross sections and other data for a CF$_4$-based plasma are listed here, partly for their own worth and partly to illustrate the processes that must be considered in setting up a plasma processing model.

First, we note that chemistry models have been employed since the outset of plasma etching; a sample, which emphasizes CF$_4$ and related gases etching Si and SiO$_2$, includes Refs. [116]–[125].

Cross sections for electrons colliding with CF$_4$ were given by:

[126] – total cross section.
[127, 128] – differential cross section.
[129] – elastic differential cross section.
[130] – vibrational excitation cross section.
[131] – total dissociation cross section.
[132, 133] – ionization cross section.
[134, 135] – double ionization cross section.

Other CF$_4$ cross sections were given by:

[136] – neutral dissociation cross section.
[137] – electron attachment cross section.
[138] – attachment and ionization attachment cross section.
[139] – electron ionization, dissociative ionization, and excitation cross sections for CF$_3$ and ionization cross section for F.
[140] – neutral dissociation cross section for F$_2$.
[141] – dissociative ionization cross section for F$_2$.
[142] – attachment cross section for F$_2$.
[143] – electron detachment cross section for F$^-$.
[144] – electron ionization cross section (discusses estimation of this cross section)
[145] – cross section for F$^-$ detachment by neutral.
[146] – cross section for dissociation of CF$_4$ by ions Ar$^+$, Ne$^+$, and He$^+$.
[92] – Langevin cross section.
[147] – elastic and momentum transfer cross sections for CF$_4$ and F.

Other data can be found in:

[124, 118, 123, 120] – neutral–neutral reaction rates.
[148] – sticking coefficients for F, CF$_2$, and CF on anodized Al.

5.6.1 Processes in the CF_4 Plasma

As soon as we estimate the rate of destruction of CF_4 by electrons, we are forced to deal with the complex nature of this discharge. CF_4 is destroyed very rapidly, and the gas phase processes that produce CF_4 are slow, yet experiments show significant amounts of CF_4 far downstream of the CF_4 inlet. These observations suggest that CF_4 might be being reformed on the only other available location – the walls. A picture will emerge, as we go into more detail, of a recycling process where molecules are broken up in the gas phase and flung to the walls. There, some of them are reformed and returned to the gas [93, 94, 92, 149].

To make progress we need to use some measured quantities. In ECR etchers the electron density n_e can be as high as or greater than 10^{12} cm^{-3}. The electron distribution is often biMaxwellian. A typical distribution might have 80% of the electrons at a temperature $T_1 = 2$ to 3 eV and the remaining 20% at a temperature $T_2 = 10$ to 15 eV. The measurements of the distribution are not very reliable in the high-energy tail of the distribution, since the number of electrons with high energies is small, making them hard to detect. The tail can be depleted by inelastic collisions, such as ionizing collisions, which remove the electron's kinetic energy, and by loss of the tail electrons to the walls. Electrons at first reach the wall faster than ions and charge it negatively until the potential builds up to repel them. Electrons with high kinetic energy may be able to overcome the potential energy barrier, which pushes them away from the negatively charged wall. Thus when electrons do escape to the wall it is from the tail that they are removed. Because the tail is responsible for driving the ionization and many of the chemical reactions, this deviation from a Maxwellian distribution may be significant.

The CF_4 feed gas is inlet into the ECR system at a rate of the order of tens of SCCM (Standard Cubic Centimeters per Minute) leading to typical gas pressures of 1–10 mTorr in etching applications. With the information given above as to the electron and neutral populations we can begin to estimate some of the most critical quantities for understanding the gas phase processes, which are the mean free paths of electrons and neutrals, for the main processes they will undergo. Particles of a given type may have several different mean free paths, one for each different type of collision. For instance, the coldest electrons can have elastic collisions with neutrals or with other electrons. The *mfp* for electrons having elastic collisions with neutrals is typically the shortest *mfp* for the electrons in an etching system. The electron–electron collision *mfp* is typically very large indeed. Hot electrons may also have inelastic collisions with neutrals and the *mfp* for these collisions will be considerably longer than the electron *mfp* for elastic collisions with neutrals.

To very roughly estimate a cross section presented by a neutral atom, suppose it has a radius of 1–2 Å, which is 1–2×10^{-10} m. The area this atom presents when looked at from the side is then $\sigma = \pi r^2$ which is from about 3×10^{-20} to 10^{-19} m^2. To convert this to a mean free path we need to know the density of neutral particles. At room temperature, 1 mTorr corresponds to about 3×10^{13} particles per cubic centimeter or 3×10^{19} m^{-3}. We would then expect typical elastic *mfp*s of $\lambda = (N\sigma)^{-1} \simeq 0.3$ to 1 m at 1 mTorr. These are as big or bigger than the dimensions of most reactors. Inelastic electron collisions have smaller cross sections and larger *mfp*s by about a factor of ten. At 10 mTorr all the *mfp*s are ten times smaller, being from 3 cm to 10 cm for the elastic processes. These are very rough estimates but they are indicative of the numbers we should expect – and these mean free paths are some of the most important parameters we need to know, to build up our understanding of the reactor.

One immediate conclusion we could draw is that if the ECR system is a cylinder with a radius of about $a = 5$ cm, neutral particles leaving the wall at high pressure have at most a few elastic *mfp*s to travel until they reach the wall again, and possibly less than one elastic *mfp* at low pressure. The *mfp* for a reaction of any type is likely to be greater than the elastic *mfp* and hence neutrals can probably cross the cylinder with little chance of a reaction.

Suppose particles of species A had a *mfp* λ_A for conversion to another species and λ_A was much smaller than the radius a and, further, suppose these particles were formed only on the walls. Then the density of species A would drop rapidly with distance from the wall. In the opposite case where $\lambda_A \gg a$ the density of neutral particles would vary only a little with distance from the wall (except perhaps near some object like the wafer, which could strongly remove particles). Thus in our etcher we expect the neutral density to depend on z (the distance along the cylinder axis from the wafer) but not on r (the radial distance from the axis) except perhaps near the wafer.

The *mfp*s we have estimated are for electrons hitting neutrals and for neutrals hitting neutrals. To describe the chemistry we also need information on the rate at which neutrals are broken up by the electrons. The electron *mfp* for inelastic collisions with neutrals is λ_e^{inel}, the electron and neutral number densities are n_e and N respectively, the density of electrons energetic enough to have inelastic collisions is n_e^*, and the electron speed is v_e. Then the frequency with which each (sufficiently hot) electron has inelastic collisions with neutrals is $v_e^{\text{inel}} = v_e/\lambda_e^{\text{inel}} = v_e N \sigma^{\text{inel}}$. The total rate per unit volume of inelastic collisions between electrons and neutrals is the rate per electron multiplied by n_e^*, the density of energetic electrons with enough energy to cause the collision to occur:

$$n_e^* v_e^{\text{inel}} = n_e^* v_e N \sigma^{\text{inel}} . \tag{5.11}$$

The rate at which each neutral has inelastic collisions is this total rate divided by the density of neutrals, or $n_e^* \sigma^{\text{inel}} v_e$.

The electron speed is about $v_e = 10^6$ ms^{-1}. One electron-volt is 1.6×10^{-19} J, and so using $v_e = 10^6$ ms^{-1}, $\frac{1}{2}mv_e^2 = 5 \times 10^{-19}$ J corresponds to about 3 eV; so this value is v_e is reasonable. We assumed σ was about 10^{-19} m^2 for elastic collisions. Suppose $\sigma^{\text{inel}} = 10^{-20}$ m^2. The density of electrons above the kinetic energy threshold, n_e^*, is probably the most difficult part of this rate to estimate. The total electron density is $n_e \simeq 10^{13}$ cm$^{-3} = 10^{19}$ m^{-3}. If we assume the threshold for the inelastic process is at the energy $\epsilon^* \simeq \alpha k T_e$ and the fraction of electrons with energy ϵ^* is $f^* = \exp(-\epsilon^*/k_B T_e)$, then $f^* = e^{-\alpha}$. If $\alpha \sim 3$ then $f^* \sim 0.03$ and $n_e^* = f^* n_e \simeq 3 \times 10^{17}$ m^{-3}. Finally, $n_e^* \sigma^{\text{inel}} v_e \simeq 3 \times 10^3$ s^{-1} is the reciprocal of the time it takes the neutral to react; $\tau_{nr} \simeq 3 \times 10^{-4}$ s.

The speed of the neutral is typically about 300 ms^{-1}. The kinetic energy associated with this speed, given a mass of very roughly 10^{-25} kg, is $\frac{1}{2}mv^2 = 5 \times 10^{-21}$ J or about $1/30$ eV, which is approximately room temperature. In the 3×10^{-4} s before the neutral reacts it can be expected to travel a total distance of 0.1 m. (Because of elastic collisions, it will go a smaller net distance.)

In the ECR system we should for instance expect CF_4 to be broken up by electrons in the first 0.1 meters, according to this estimate. More detailed calculations of gas phase rates confirm this picture and, what is more, they do not suggest a gas phase process to replace CF_4. This, and experiments that show considerable amounts of CF_4 throughout the reactor, provide some reason to suppose CF_4 is being formed on the reactor walls and being released into the gas phase. The most detailed modeling studies of the transport of neutrals and charged species in a CF_4 plasma in an ECR system, which allow for the finite particle mean free path and other nonlocal transport processes, were done by Harvey [92–94], who assumed that CF_x recycled off the walls in some fashion. Miyata [149] observed this recycling, as a result of the sputtering of the fluorocarbon film that tends to form on the walls of the system.

Exercises

1. Monomers of a gas coalesce when they collide with other monomers, or larger "dust particles" composed of multiple monomers. The number of monomers in a dust particle is m. The number per unit volume of dust particles of size m is N_m. Suppose that the dominant mechanism for dust particles to grow is by collecting monomers. Then the rate of increase of the density N_m of dust particles with m monomers is proportional to the density N_{m-1} of dust particles with $m - 1$ monomers, multiplied by the density of (free) monomers, N_1. If N_1 is constant, and dust particles do not lose monomers, then

$$\frac{dN_m}{dt} = \alpha_{m-1}N_{m-1} - \alpha_m N_m. \qquad (5.12)$$

For more explanation of this equation, see Ref. [150].

The positive term on the right-hand side represents production of dust particles of size m, when clusters of size $m - 1$ capture a monomer. The negative term represents loss of dust particles of size m, when they grow to size $m + 1$ by capture of a monomer.

This equation can be rewritten using an "integrating factor" method as

$$\frac{dN_m}{dt} + \alpha_m N_m = e^{-\alpha_m t}\frac{d}{dt}(N_m e^{\alpha_m t}) = \alpha_{m-1}N_{m-1}, \qquad (5.13)$$

so that

$$\frac{d}{dt}(N_m e^{\alpha_m t}) = \alpha_{m-1}N_{m-1}e^{\alpha_m t}. \qquad (5.14)$$

Assuming N_1, the density of monomers, is constant and that N_2 obeys an equation of the above form, we find

$$\frac{d}{dt}(N_2 e^{\alpha_2 t}) = \alpha_1 N_1 e^{\alpha_2 t}. \qquad (5.15)$$

Integrating this, we get

$$N_2 e^{\alpha_2 t} = A + \frac{\alpha_1 N_1}{\alpha_2} e^{\alpha_2 t}, \tag{5.16}$$

and so

$$N_2 = A e^{\alpha_2 t} + \alpha_1 N_1 / \alpha_2. \tag{5.17}$$

If N_2 was initially zero, then $A = -\alpha_1 N_1 / \alpha_2$ and

$$N_2 = \frac{\alpha_1 N_1}{\alpha_2} (1 - e^{-\alpha_2 t}). \tag{5.18}$$

a) Calculate N_3, assuming $\alpha_3 = \alpha_2$.

b) Calculate N_3 and the next few N_m values, assuming $\alpha_m > \alpha_{m-1}$. Write a general expression for the mth dust particle size, N_m, in terms of a sum over terms corresponding to all the lower m values.

c) Assuming dust particles of size 10 drop under gravity out of the gas before they have time to grow (this assumption is not justified in reality), plot the rate of arrival of dust at the bottom of the chamber, as a function of time. Let $\alpha_m = \alpha_1 (1 + m/20)$, and use $\tau_1 \equiv \alpha_1^{-1}$ as one unit of time in the plot.

Show the variation until steady state is approached.

2. A gas x undergoes reactions:

$$\begin{aligned} x + e &\to x^+ + 2e \quad \text{(ionization)}, \\ x + e &\to x^- \qquad\quad \text{(attachment)}, \\ x^- + e &\to x + 2e \quad \text{(detachment)}. \end{aligned}$$

The rate of ionization per unit volume is $R_i = r_i(T_e) n_x n_e$, where T_e is the electron temperature in eV, n_x is the number density of species x, and n_e is the electron number density. The rate of attachment is $R_a = r_a(T_e) n_x n_e$. The rate of detachment is $R_d = r_d(T_e) n_{x^-} n_e$, where n_{x^-} is the number density of the x^- ion. This question is concerned with the relative densities of x^+ and x^- ions. The x^+ ions are lost from the plasma, because the electric field pulls them to the wall. The x^- ions are only removed by detachment – the electric field holds them in the interior of the plasma. Attachment is usually faster at low temperature T_e, whereas detachment is faster at high T_e. Let $r_i = 10^{-12} T_e \, \mathrm{cm}^{-3} \mathrm{s}^{-1}$, $r_a = 10^{-16} / T_e \, \mathrm{cm}^{-3} \mathrm{s}^{-1}$, and $r_d = 10^{-12} T_e \, \mathrm{cm}^{-3} \mathrm{s}^{-1}$. Assume x^+ ions have a mean velocity of $10^4 \sqrt{T_e}$ cm s^{-1} and that they have to travel a mean distance of 10 cm to escape; thus $\tau_+ = 10^{-3} T_e^{-1/2}$.

If $n_x = 10^{16}$ cm^{-3}, what are n_{x^-} and n_{x^+} if $T_e = 1$ eV? Set $n_{x^+} = n_e + n_{x^-}$, for charge neutrality. What is n_{x^-} / n_{x^+}? Repeat the calculation for $T_e = 4$ eV. Suggest more realistic temperature dependences for $r_i(T_e)$, $r_a(T_e)$, and $r_d(T_e)$.

6

Transport at Long Mean Free Path

In this chapter we examine long mean free path transport. This topic will involve reconsidering some of the chemistry issues from Chapter 5 and will also introduce material relevant to trench profile evolution, which is handled primarily in Chapter 7. We begin with long mean free path transport and its effect on the chemistry of neutrals in a cylindrical discharge. Ion transport and etching by ions are introduced; this is followed by a discussion of long mean free path transport in a trench that is being etched.

6.1 Chemistry at Long Mean Free Path

When the mean free path λ is large compared to the reactor dimension L (which we take to be a cylinder radius a) the particles have collisions every time they travel a distance equal to about $2a$, the cylinder diameter. If $\lambda > L$, the particles predominantly have collisions with the walls. If they are moving mainly in the z direction they may go further than this, but if they are moving nearly along the surface of the wall with little motion in the z direction they will travel far less. If the mean free path (i.e., the distance traveled between collisions, including collisions with the wall) is about $2a$ then the diffusion coefficient is

$$D \simeq \frac{1}{2}v\lambda = \frac{1}{2}v2a = va. \tag{6.1}$$

In this formula, the step size (which is normally taken to be λ) was set equal to $2a$. The long mean free path motion of neutrals down the length of the cylinder will be diffusive, with this diffusion coefficient, provided the neutrals are not likely to react each time they hit the wall. In our example where $a = 5$ cm we get $D \simeq (300)(0.05) = 15 \ \mathrm{m^2 s^{-1}}$.

If the cylinder is about 50 cm long we could estimate the time to diffuse from the inlet to the outlet using this value of D. The distance diffused in a time t is about $L \simeq \sqrt{2Dt}$. To go a distance L will take

$$t \simeq L^2/(2D). \tag{6.2}$$

In our case, $L = 0.5$ m and $t = (0.5)^2/30 \simeq 10^{-2}$ s. In this time the particle travels a total distance (backwards and forwards) of about $vt = 3$ m. An equivalent way of looking at this is to say that the mean distance L traveled, in a random walk in N steps of length λ, is

$$L \sim \sqrt{N} \lambda. \tag{6.3}$$

Then $N = (L/\lambda)^2$ and if for the step size λ we use $\lambda = 2a$ with $a = 5$ cm, then the number of steps is $N = 25$. The total distance traveled is $N\lambda$ or 2.5 m. The significance of this is that to go the 0.5 m down the tube, particles in this case go a total distance about five times greater, taking longer than if they did not hit the walls and having a five times greater chance of chemical reaction or other collisions.

6.1.1 Rates for CF₄

Using this information about the time it should take to diffuse down the cylinder, and the reaction rates for CF_4 [94], we can examine the chemical reactions expected and estimate the neutral densities. First we need to consider the effect the electrons have on the CF_4 that is pumped into the chamber. CF_4 can be dissociated into neutral products losing one, two, or three fluorine atoms with significant probabilities. (This problem has been addressed computationally, [93, 94] using the long mean free path method described in Section 6.6, where it is used to handle neutral particles' transport in a trench.)

The overall number of dissociations, per unit volume per second, is

$$R_{\text{diss}} = r_{\text{diss}} N_{CF_4} n_e. \tag{6.4}$$

The dissociation rate depends on the electron distribution assumed, but a typical combined rate coefficient for dissociation into neutrals is about $r_{\text{diss}} = 2 \times 10^{-9}$ cm³s⁻¹. The combination of $r_{\text{diss}} n_e$ is equal to the integral given in Equation 5.8, the rate of reaction per neutral molecule, (which could also be called the frequency of reaction for the neutral molecules) ν_{diss} which is $\nu_{\text{diss}} = \int_o^\infty \sigma^{\text{diss}} v_e f d\epsilon$. Since the electron distribution function f is proportional to n_e we can factor out n_e and define $r_{\text{diss}} = \nu_{\text{diss}}/n_e$. The total rate of dissociation R_{diss} is equal to the rate (or frequency) of dissociation of each CF_4 molecule ν_{diss} multiplied by the number density of CF_4,

$$R_{\text{diss}} = N_{CF_4} \nu_{\text{diss}}, \tag{6.5}$$

where $\nu_{\text{diss}} = r_{\text{diss}} n_e$. The measured electron density is about 10^{12} cm⁻³ and so $\nu_{\text{diss}} \simeq 10^{12} r_{\text{diss}} = 2 \times 10^3$ s⁻¹. In other words, the time to dissociate, $\tau_{\text{diss}} \equiv \nu_{\text{diss}}^{-1} \simeq 0.5$ ms. This is much less than the time to diffuse down the cylinder.

The rate of ionization of CF_4 depends rather more on the details of the electron distribution, because very energetic electrons are needed to do the ionization. For one reasonable assumption about the electron distribution, the total CF_4 ionization rate

$$R_{\text{ioniz}} = r_{\text{ioniz}} N_{CF_4} n_e \tag{6.6}$$

has a rate coefficient r_{ioniz} that is similar to r_{diss} quoted above; $r_{\text{ioniz}} \simeq 2 \times 10^{-9}$ cm³s⁻¹. The frequency of ionization of each CF_4 molecule is $\nu_{\text{ioniz}} = r_{\text{ioniz}} n_e \simeq 2 \times 10^3$ s⁻¹. The total rate of removal of CF_4 is the sum of these rates and the lifetime of the CF_4 is $\tau = 1/(\nu_{\text{diss}} + \nu_{\text{ioniz}}) = 0.25$ ms.

The density of CF_4 could be replenished by other processes involving reaction of the neutrals in the gas phase. The most likely candidates, according to Refs. [92]–[94], are

$$CF_3 + F_2 \rightarrow CF_4 + F, \tag{6.7}$$

which has a rate coefficient of $r_1 = 7 \times 10^{-20}$ cm^3s^{-1}, and

$$CF_3 + F \rightarrow CF_4, \tag{6.8}$$

with rate coefficient $r_2 = 1.1 \times 10^{-19}$ cm^3s^{-1}, at a pressure of 1mTorr. These rate coefficients are much smaller than the rates for electron reactions but the densities of neutral species are much higher. As mentioned above, the total neutral density at 1 mTorr should be about 3×10^{13} cm^{-3}, so if we take a typical neutral species density to be 10^{13} cm^{-3} (an overestimate) the rate of reaction of CF_3 with F_2 would be

$$R_1 = r_1 N_{CF_3} N_{F_2} = 7 \times 10^{-20} (10^{13})^2 \tag{6.9}$$

and the rate of reaction of CF_3 with F would be

$$R_2 = r_2 N_{CF_3} N_F = 1.1 \times 10^{-19} (10^{13})^2. \tag{6.10}$$

The total production rate of CF_4 is the sum of these, or 1.8×10^7 s^{-1}. This is a very small production rate, despite our using overestimates of the densities of F_2 and F.

The removal rate of CF_4 by electron processes, found above, was

$$\frac{N_{CF_4}}{\tau} = N_{CF_4}(\nu_{diss} + \nu_{ioniz}), \tag{6.11}$$

where τ is the lifetime. If any species is removed at a rate N/τ per unit volume and produced at a rate S per unit volume (and if we ignore other ways of providing new particles, such as them flowing into the region) then in steady state these rates must be equal:

$$S = \frac{N}{\tau} \quad \text{or} \quad N = S\tau. \tag{6.12}$$

The production rate of CF_4 is about $S = 1.8 \times 10^7$ s^{-1}, and the lifetime is at most $\tau = 2.5 \times 10^{-4}$ s; therefore, the two processes we considered could provide a density of $S\tau \simeq 4.5 \times 10^3$ cm^{-3}, which is extremely small. From this we conclude that production of CF_4 in the gas phase is unimportant at this pressure. This is despite the fact that, in experiments, 10–20% of the neutral gas in the downstream region is CF_4. The only explanation we know of for this discrepancy is that CF_4 must form on the chamber walls (Fig. 6.1).

The problem with this explanation is that we know very little about processes on the walls. There is some experimental evidence that a small fraction γ of the particles that hit the wall will stick to the wall; γ is of the order of a few percent. What happens to the particles stuck to the surface is not known. We will assume that the particles will reform CF_4 as much as possible. If there is more F than is needed to form CF_4 then we assume CF_4 and F_2 come off the surface. If there is too

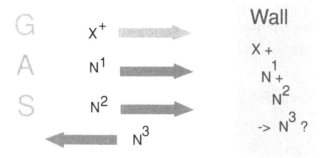

Fig. 6.1. Recycling of gases off the chamber walls. Positive ions, denoted X^+, and neutral particles denoted N^1 and N^2, strike the wall. The ions are neutralized when they contact the wall. Neutral species N^3 is produced at the wall and enters the gas phase. (X and N stand for unspecified species.) The chemistry of the reactions taking place on the wall has a profound effect on the composition of the gases recycling off the wall.

little F for the available carbon we assume that as much CF_4 as possible is formed as well as some CF_3. This seems plausible but it could be significantly wrong.

6.1.2 Rates at a Surface

If we believe that all the CF_3 molecules that stick can pick up an F atom and form CF_4 we can estimate densities rather simply. The molecule only sticks on average once every $1/\gamma$ times it hits the surface. If it travels on average the cylinder diameter of $2a$ between successive impacts on the surface then it travels a distance of $2a/\gamma$ on average before sticking, which takes $\tau = 2a/\gamma v$ seconds. For CF_3 this would be about $\tau = 10^{-4}/\gamma$ or, for $\gamma = 0.1$, $\tau = 1$ ms.

We estimated above that the lifetime of a CF_4 molecule is limited by gas phase reactions to a value of about $\tau_{CF_4} \simeq 2.5 \times 10^{-4}$ s. Our estimate of the time needed for CF_3 to have a wall reaction gave a similar result, to within the accuracy of our estimates. If the rate of breaking up CF_3 in the gas phase is similar to that for CF_4, then we have shown that the CF_3 lifetime due to gas-phase reactions may be comparable to the time for the CF_3 reaction at the wall. In addition, CF_3 is also broken up into other fractions in the gas phase, most of which have the carbon atom in a positive ion. Because the rate for CF_3 being ionized is similar to the rate for CF_4 being broken up, this process gives a lifetime for CF_3 of about $\tau \simeq 0.25$ ms. But the ions that are formed go to the surface very quickly under the influence of the electric field, where they are more likely to stick than neutrals because they are moving rapidly toward the surface. If all the ions stick, they will then react on the wall. The time before CF_3 reacts on the wall would be set by the time for CF_3 to be ionized in the gas phase, if this time is shorter than the time for a neutral to stick to the surface.

In summary, CF_4 is being turned into CF_3 in the gas phase and is being turned back again on the wall, with each species having similar lifetimes and thus also similar densities. Changing the assumptions can change this conclusion dramatically, however.

To use the "surface model" accurately we need to know the flux of particles striking the surface. If the particles were moving with speed v toward the surface, then in one second they would move v meters, and so all the particles within v meters of the surface would hit the surface in a second. Above a unit area of surface, a volume v meters high and one square meter in cross section or v m^3 in volume will hit the surface in a second. The number of particles in this volume is the number density N multiplied by the volume, and the volume is equal in magnitude to the speed v. The number hitting the unit area of surface per second is the flux per unit area; therefore, the flux is per unit area is $\Gamma = Nv$. If the particles are moving in random directions the number striking per unit area per second is smaller:

$$\Gamma = \frac{1}{4}Nv. \tag{6.13}$$

The number attaching to a unit area per second is the fraction that sticks, γ, times the flux per unit area. For CF$_3$ if $v \sim 300\,\mathrm{ms}^{-1}$ and $N \sim 10^{19}\,\mathrm{m}^{-3}$ then if $\gamma = 0.05$ the number sticking to a square meter per second is about $4 \times 10^{19}\,\mathrm{m}^{-2}\mathrm{s}^{-1}$ or $4 \times 10^{15}\,\mathrm{cm}^{-2}\mathrm{s}^{-1}$. If these all pick up an extra F then this same number of CF$_4$ molecules leave the surface per unit area per second.

If the sticking coefficient γ_i for ions were less than one, then the fragments striking to the wall but not sticking might be neutralized and directly recycled into the gas phase, without having the opportunity to react on the surface. The same type of arguments can be used to estimate production rates and lifetimes for all the other neutral species, all of which have the same high degree of uncertainty associated primarily with the surface processes.

The argument used here assumed the estimates could be done ignoring gas flow down the tube. The time to cycle from the gas phase to the wall and back into the gas phase is much shorter than the time to diffuse down the cylinder, as was shown by the calculations. The local recycling equilibrium should be roughly the same all the way down the cylinder because the conditions at the input are rapidly "forgotten" and the equilibrium values are quickly established. However, if the electron density is much higher at some points, that higher electron density will shift the equilibrium of the neutral particles we described, by changing the rates of the electron processes.

A more detailed calculation than this can explain some features that are seen experimentally, but it suffers from exactly the same primary difficulty that quantities such as the sticking coefficient γ are not known. To address this problem we must examine suitable experiments, which could test the overall form of the models used and establish values for quantities such as γ. Before we examine the appropriate experiments we will describe in more detail the ion chemistry taking place in this CF$_4$ discharge.

6.1.3 Ion Chemistry

One interesting experimental observation in the CF$_4$ discharge in the ECR reactor is that the neutral density sometimes exhibits a hollow profile. This is attributed to "ion pumping". Ion pumping refers to a process where ions, which are moving out of the plasma, collide with neutrals and push them away from the centre of the plasma. The positive ions are pushed out of the center of the plasma by the electric field. If such an ion has a "charge-exchange" collision, it picks up an electron from a neutral. The new neutral, which was previously a fast-moving ion, has a high velocity directed out of the plasma. The fast-moving neutral tends to leave the center of the plasma and the slow ion begins to accelerate out of the plasma under the influence of the electric field. If the mean free path for charge exchange is λ_i^{cx} and the distance to the wall from the point where the ion is first created is L, then the ion leads to roughly L/λ_i^{cx} charge-exchange collisions. Each charge exchange collision leaves a fast neutral heading for the wall and creates a new ion, which will move toward the wall. Each positive ion created by electron impact was made from an original neutral, but it leads (because of charge exchange) to an additional L/λ_i^{cx} neutrals being launched toward the wall. If $L \gg \lambda_i^{cx}$ then the charge-exchange process removes many more neutrals from the gas phase than electron impact ionization does directly. (Many of these neutrals will bounce off the walls and return to the gas phase, however.)

The hollow density profile in the ECR system also depends on the fact that ions tend in part to follow magnetic field lines. If the ions (and fast neutrals they create through charge-exchange) went straight to the nearest wall the density would probably not hollow out. Because the field lines partially direct the ions to the ends of the system, when they are recycled into the gas phase as neutrals they produce extra neutrals and a higher density at the ends of the reactor. This is the origin of the hollow density profile.

6.2 Ion Behavior Near the Surface and Evolution of Surface Features

How the shape of a semiconductor surface changes during processing of the surface is one of the critical issues in semiconductor fabrication. Even when the surface is flat and film is being deposited, very complex structures can evolve. The most common situation where a precise shape is wanted, however, involves etching or deposition inside a trench.

To predict the way the shape evolves it is necessary to know how the surface is affected by the particle fluxes, but the information available is, again, incomplete. Models for etching rates have been proposed and will be discussed next along with film deposition. The calculation of how the surface evolves will be described in detail and illustrated using approximate models in Chapter 7. Where possible, the expressions for etch and deposition rates should be estimated based on experimental data.

Plasma etchers are able to provide etching that is much more anisotropic than can be obtained with purely chemical etchers. The standard interpretation of this observation involves the fact that, for some materials and etch gases, a combination of reactive neutral species and energetic ions are necessary to achieve rapid etching. There are several similar reasons why a combination of neutrals and ions might be needed. These explanations all assume that the ions somehow activate the surface and the neutrals follow up on this activation by chemically reacting with the surface.

Which explanation applies probably depends on the nature of the surface and on the chemical mix in the gas phase. A straightforward example occurs when the gas input to the system is CF_4 since, in some circumstances, CF_4 will form a tough polymer that deposits on the surface. This polymer protects the surface from chemical attack, unless ion bombardment (or chemical reactions) removes the polymer from the surface.

The directionality or anisotropy of the etch is a result of the need for energetic ions to "activate" the surface and the fact that the ions are accelerated in the sheath. The sheath near the surface is a region of strong electric fields. The electric fields are set up by the mobile electrons in the plasma, which initially charge up the surface faster than positive ions can reach the surface. The negative charge on the surface eventually builds up until it repels electrons and attracts positive ions strongly enough to make electrons and positive ions arrive at equal rates. The electric field lines start on positive charges in the sheath and end on negative charges on the surface. The sheath is predominantly positive, because the negative charge on the surface repels plasma electrons, leaving behind the ions that are "falling" out of the plasma.

The sheath must have enough electrostatic potential across it to repel all but the fastest moving electrons. If we assume the electrons have a temperature T_e measured in eV then the sheath voltage in volts should be about 3–5 T_e. The ions fall through the sheath and pick up this much energy. Provided they do not have any collisions while they are crossing the sheath, the extra velocity they acquire is (essentially) all in the direction toward the wall. Because they are moving toward the wall of the reactor they are likely to hit the bottom of a trench but not the sidewalls of the trench. The trench bottom is activated and will be etched by the chemicals but the sidewalls will not, leading to directional (downward) etching.

If the size of the trench is typically a micron (10^{-6} m), the sheath will usually be much larger than the trench. This follows from Poisson's equation,

$$\frac{d^2V}{dx^2} = -\rho/\epsilon.$$

The ion density n_{is} in the sheath is at most about a tenth of the maximum ion density, which is not usually more than 10^{12} cm^{-3} or 10^{18} m^{-3}; so $n_{is} \simeq 10^{17}$ m^{-3}. There is essentially no electron density in the sheath (the fast electrons that cross the sheath contribute very little density). Taking n_{is} to be constant in the sheath (which means we have a so-called matrix sheath) then

$$\frac{d^2V}{dx^2} = -n_{is}e/\epsilon_o.$$

Integrating gives

$$\frac{dV}{dx} = (n_{is}e/\epsilon_o)(x_s - x),$$

where x is the distance from the wall at $x = 0$ and x_s is the sheath thickness. Finally,

$$V = -(n_{is}e/2\epsilon_o)(x_s - x)^2.$$

(The derivation assumes the wall is to the left and the result only applies inside the sheath.) If the maximum potential drop is $V \sim 5T_e$ (with T_e expressed in eV) then the sheath thickness is given by

$$5T_e \sim \left(\frac{n_{is}e}{2\epsilon_o}\right)x_s^2. \tag{6.14}$$

Thus $x_s \simeq (10\,\epsilon_o\,T_e/n_{is}e)^{1/2}$ and if we approximate $\epsilon_o \simeq 10^{-11}$ F/m we have $x_s \simeq (5 \times 10^{-10}/1.6 \times 10^{-2})^{1/2} \sim (3 \times 10^{-8})^{1/2} \simeq 1.7 \times 10^{-4}$ m. This value of the sheath thickness is probably an underestimate because we took a small voltage and a high density, but it is still over a hundred times bigger than the trench size.

It is possible for the etcher to be operated at high enough neutral density for significant numbers of ion collisions with neutrals to occur as ions cross the sheath. Even when x_s is very small compared to the mean free path λ, a fraction of ions equal to x_s/λ will collide as they cross the sheath. The mean free path is defined so that the fraction of the ions that have not collided after traveling a distance ℓ obeys the equation

$$\frac{df}{d\ell} = -\frac{f}{\lambda}, \tag{6.15}$$

which has the solution $f = e^{-\ell/\lambda}$. If ℓ/λ is small, $e^{-\ell/\lambda} \simeq 1 - \ell/\lambda$ is the uncollided fraction. The fraction that has collided is then ℓ/λ. An elastic collision changes the ion energy and direction and so elastic collisions in the sheath decrease the anisotropy of the etching because they increase the chance of ions moving sideways and hitting the sidewall of the trench.

Inside the trench the chance of a collision is very small indeed, being of order the trench dimension divided by λ. There is a possibility that, if the walls or bottom of the trench are covered by an insulating layer, charge will build up there and deflect the ions. The charge builds up until it makes the net flux of charge to any point on an insulator be zero. Hence the electron charge multiplied by the potential that is set up must usually be comparable to the particle energies. In the trench both ions and electrons have energies of about T_e, because the ions fell through the sheath potential of several T_e.

These arguments have shown that the ion distribution depends on the sheath potential and may also depend on collisions in the sheath and potentials inside the trench.

6.3 Neutral Particle Kinetics at Surfaces

The distribution of neutrals in the trench is determined by processes similar to some of those affecting the ions. There are essentially no gas phase collisions of neutrals inside the trench but neutrals do collide with the walls. They are probably not very likely to stick to the wall any given time they bounce off it but will instead bounce around for some time before attaching to it. Each time they bounce, their direction of motion has a chance to be randomized. If they simply bounced off a smooth, flat surface without attaching to the surface at all, they might retain some memory of their initial direction. Roughness of the surface would alter their direction to some extent even in the case that they only bounce. At the other extreme, if the neutrals attach to the surface they will quickly "forget" their original motion and when they eventually escape they will probably produce a distribution in the gas phase that is close to being isotropic.

Before describing in detail how neutral particle behavior is calculated, some preliminary discussion will be given of the angular distribution of the neutral particles. We focus on the angular distribution in the gas phase and the angular distribution of particles leaving a surface. As we shall see, for isotropy in the gas phase, the number leaving the surface must be peaked in the direction away from the surface.

Suppose the particle distribution is isotropic. First we construct a sphere, as shown in Fig. 6.2, and consider the particles coming from very close to the center of the sphere. Equal numbers of particles will go through equal areas of the surface of the sphere. If the sphere is close to a surface it is useful to let $\hat{n}$ be the unit vector normal to the surface and use a spherical coordinate system to describe the particles' velocities. The angle θ is the angle between the normal $\hat{n}$ and the direction of some particle's motion. Now suppose we can measure the number of particles leaving the surface at an angle θ to the normal $\hat{n}$ and crossing a small area dS, traveling at right angles to the area dS. If the small area dS is truly flat and the

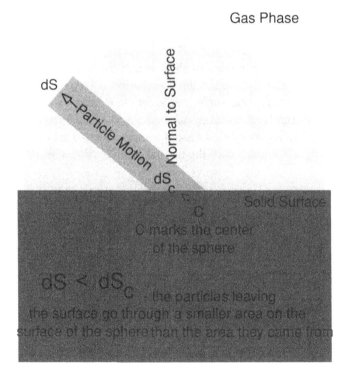

Fig. 6.2. The small area dS on the surface of the sphere is projected onto a larger area dS_c at the center of the sphere, which coincides with the solid surface.

particles cross it exactly at right angles then if we follow the particles back to the surface they will have come from an area dS_c equal to $dS/\cos\theta$. To understand this observation, suppose we look down a long narrow tube at the surface, so that only parallel rays of light could travel all the way down the tube. Then we can see a larger area of surface if θ is large than if θ is close to zero. Since the particles mentioned earlier cross the area dS at right angles, we can imagine that the area is part of the surface of a large sphere and the particles come along radii from near the center of the sphere. The distribution leaving the surface is isotropic and so the number of particles crossing dS should not depend on θ. For this to be true the number leaving a unit area on the surface at an angle θ to $\hat{n}$ must be proportional to $\cos\theta$. The area on the surface dS_c sending particles through dS, at right angles to dS, when the particles are traveling at an angle θ to the normal, is $dS_c = dS/\cos\theta$. The number leaving unit area per second at an angle θ, in the range $d\theta$, is equal to $N_o\cos\theta d\theta$, where N_o is some constant. The number leaving area $dS/\cos\theta$ on the surface and crossing dS within $d\theta$ of being at right angles to dS is then $N_o dS d\theta$, which is independent of θ and proportional to $d\theta$ as it should be for an isotropic distribution.

In our scenario above dS did not depend on θ or ϕ but if we allow dS to be the area traced out by a radius r, when θ changes by $d\theta$ and the azimuthal angle changes by $d\phi$, then $dS = r^2\sin\theta\,d\theta\,d\phi$. The solid angle $d\Omega$ subtended by dS as seen from $r = 0$ is dS/r^2 if dS is at right angles to the radius. Thus $d\Omega = \sin\theta d\theta d\phi$. Equal numbers of particles go into equal solid angles in an isotropic distribution. In the gas phase the number per second going from near a point into $d\Omega$, as seen from that point, in an isotropic distribution is proportional to $d\Omega = \sin\theta d\theta d\phi$. Under isotropic conditions the number coming off the surface per unit area per second and going into $d\Omega$ is proportional to $\cos\theta d\Omega = \cos\theta\sin\theta d\theta d\phi$.

What this discussion is intended to show is that the flux coming off the surface can have a θ dependence even when the distribution is isotropic. The flux per unit area of the surface leaving the surface at an angle θ varies as $\cos\theta$, whereas the flux leaving the surface and crossing the area $dS = r^2\sin\theta d\theta\,d\phi$ traced out by a radius as it swings through angles $d\theta$ and $d\phi$ varies as $\cos\theta\sin\theta$. This information allows us to set up a calculation of the neutral particle fluxes inside a trench.

6.4 Neutral Particle Transport in a Trench

In this section we tackle neutral transport by a variety of methods. We begin with a computational method [151, 152] and then turn to simple analytic models. The problem of neutral transport is often tackled using Monte Carlo methods [153–158]. Recently, nonstatistical methods similar to that given here have also been developed [159–165]. (A similar calculation has been used to model ion transport during ion implantation [17]).

6.4.1 Computational Approach

The motion of neutral particles as they diffuse down the trench is a clear example where ideas connected with long mean-free-path transport are needed to obtain useful information. The objective of the material in this section is to show how a simple computer calculation can be set up, which will yield the fluxes of neutral particles to each point on the sides and bottom of the trench.

The approach suggested here involves using the angular distributions discussed above to find the probability $P_{i'}^i$ that a particle, which hits one section of the boundary labeled i', will next hit another section labeled i. The boundary is divided into short sections, each of which is labeled by an integer. The rate of etching at each point on the surface can presumably be found from the particle fluxes. How the surface shape changes because of that etching is calculated in various different ways. In order to discuss the neutral particle transport in the trench, we divide the trench walls into short straight sections. Each of these short sections, labeled i, is etched at a given rate $E(i)$. Then the simplest assumption would be that each section's surface moves backward in a direction normal to the section itself. Provided the initial sections are not in the same line, the surface will behave in a physically reasonable way that is straightforward to keep track of. If adjacent sections are in the same line, then the different rates of etching of different sections lead to the creation of new sections.

In Fig. 6.3 the sections are at right angles to the sections on either side of them. Since the initial shape is broken into sections that are either parallel or at right angles, they stay that way. In other words, all the sections will always be parallel or at right angles, which simplifies calculations.

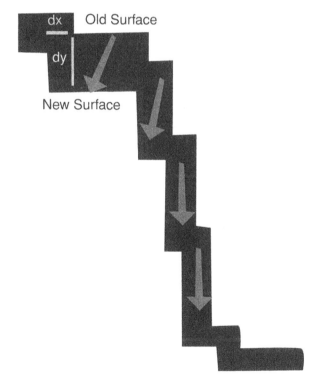

Fig. 6.3. Evolution of the sections of the surface during etching. Arrows indicate the size and direction of the etching.

Another way to describe the trench evolution is to divide the volume into small cells and allow the cells to convert from "full" to "empty" as etching progresses. This is the easiest way to get started with computation of the trench evolution, but is less accurate unless the cells are very small.

The number of particles per second entering the trench through the top and hitting each section of the wall is the input to the calculation. This is denoted $R_i^{(o)}$ to indicate the rate at which particles hit section i, the superscript o showing that this rate only includes that generation of particles that had no previous bounces and that now are scattering off the wall. This rate of scattering, of particles having their first bounce off the trench sides, is found from the angular distribution of neutrals outside the trench.

Once a particle scatters off a section labeled i', that particle has a probability $P_{i'}^i$ of scattering next off any other section i. The total rate of scattering off section i' is $R_{i'}$. A fraction of those particles equal to $P_{i'}^i$ scatters next off section i. Therefore, in steady state, the rate of scattering off i, of particles having immediately beforehand scattered off i', is $P_{i'}^i R_{i'}$. The number scattering per second off i, which came off from all other cells i', is the sum of this over all i'. The only other contribution to the number per second scattering off i is from the particles that entered the trench from outside without bouncing yet, and this rate is given by $R_i^{(o)}$. Combining these, we get the total rate of scattering off section i, in steady state:

$$R_i = R_i^{(o)} + \sum P_{i'}^i R_{i'}. \qquad (6.16)$$

Particles reaching the top of the trench are lost.

For example, suppose that section $i = 10$ is hit by 10^{16} particles per second, which just came through the top of the trench and have not yet hit the trench sides. This means that $R_{10}^{(0)} = 10^{16}$ s^{-1}. Suppose that particles that hit section 100 have a 1% chance of next hitting section 10; so $P_{100}^{10} = 0.01$. If 10^{17} particles per second in total hit section 100, then $R_{100} = 10^{17}$ s^{-1} and section 100 contributes $P_{100}^{10} R_{100} = 0.01(10^{17})$ particles per second, which hit section 10. Every other section also contributes to the flux in section 10 and so the sum runs over all the sections:

$$R_{10} = R_{10}^{(0)} + P_{100}^{10} R_{100} + P_{99}^{10} R_{99} + P_{98}^{10} R_{98} + \cdots = 10^{16} + 0.01(10^{17}) + \cdots.$$
$$(6.17)$$

If particles can stick to the surface then we shall say that they will do so with a probability γ. The probability of going to any other section of the surface, after hitting the surface, is reduced to a fraction $(1 - \gamma)$ of what it would be if there were no sticking. If the probability $P_{i'}^i$ were calculated ignoring sticking, then the actual probability would be $P_{i'}^i(\text{actual}) = (1 - \gamma) P_{i'}^i$ and

$$R_i = R_i^{(o)} + \sum_{i'} (1 - \gamma) P_{i'}^i R_{i'}. \qquad (6.18)$$

If we know the first-generation scattering rate $R_i^{(o)}$, the sticking coefficient γ, and the probabilities $P_{i'}^i$ then we can iterate this equation to find the rates R_i. This means we take an initial guess at the set of R_i (probably simply using $R_i^{(o)}$) and use this formula, with the initial guesses at the R_i inserted on the right-hand side,

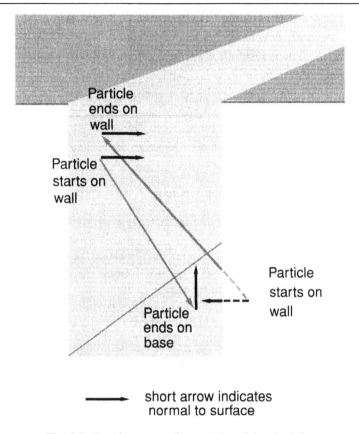

Fig. 6.4. Trench geometry for neutral particle calculation.

to find a better guess. Then we use this improved guess on the right-hand side to find an even better guess, and so on. This procedure converges quickly to give the steady-state scattering rates R_i.

We now turn to the main difficulty in this approach: finding the probabilities. To show how the probabilities are found, we begin with a trench, as shown in Fig. 6.4. We will approximate the probabilities at first. The approximations can be improved by integrating carefully over the initial and final cells, since at first we evaluate all quantities at the cell centers. In either case we impose the normalization condition

$$\sum_i P_{i'}^i = 1 \tag{6.19}$$

for each initial cell i', to ensure that particles leaving the initial section i' have a probability of exactly one of going somewhere. (For this to make sense the top of the trench must be included in the sum over final sections i). The probabilities must be normalized accurately, so that the particles leaving each cell are conserved. Particles reaching the top are lost as far as the main calculation is concerned.

The probability we shall calculate first is for movement between opposite walls. Suppose the wall is divided in the vertical coordinate y into sections of equal width Δy. The position y at the center of the jth cell is $(j - 1/2)\Delta y$. Suppose that the sections are also divided in the direction along the trench, z, into sections of width Δz. The distance in y between the centers of the sections we consider is $(j - j')\Delta y$. The separation in z is $m\Delta z$, where m is some integer. The total distance r between the cell centers is given by

$$r^2 = w_t^2 + (j - j')^2(\Delta y)^2 + m^2(\Delta z)^2 \tag{6.20}$$

where w_t is the trench width.

The trajectory of the particle subtends the same angle θ to both surfaces, when the surfaces are parallel, given by

$$\cos\theta = w_t/r. \tag{6.21}$$

If the particle went between the side and the bottom of the trench, the angles would not be equal.

The area of the final section is $\Delta s_f = \Delta y \Delta z$. From the center of the initial section the projected area of the final section is roughly $\Delta s_f \cos\theta$, and the solid angle $\Delta\Omega_f$ it subtends is this area divided by r^2:

$$\Delta\Omega_f = \Delta y \Delta z \cos\theta/r^2. \tag{6.22}$$

The probability of a particle from the initial section going into this solid angle is proportional to the solid angle. Because the flux coming off a surface isotropically also varies as $\cos\theta$, the flux reaching the final section, from a given initial section, is proportional to $\cos\theta\,\Delta\Omega_f$. The constant of proportionality is found by summing these probabilities over all final sections, including the top and bottom of the trench, and setting the total to one.

The probability of going between a cell on the wall and a cell on the trench bottom is found similarly [151].

This approach can be extended to include the effects of collisions in the gas phase. If gas phase collisions are included then the interior volume must first be divided into cells. The probabilities must link not only all possible combinations of initial and final sections of the surface but all possible initial and final volume cells. Finally, they must also allow for particles starting on the surface and scattering in the volume and vice versa. Gas phase collisions are not likely to be important in a trench of typical size. However, they are important in the main plasma.

Similar formulations of the neutral transport in a trench have been used to predict the evolution of a trench profile. The shape of the trench at various times during etching is in excellent qualitative agreement with experiment. It is difficult to establish quantitative agreement because the extent to which the particles are isotropized when they hit the walls and the sticking coefficient γ are not known. The rate of etching under different conditions is also not well known. If γ is determined by comparison with one set of experimental data then excellent extrapolations can be obtained. Equipped with a sophisticated model of the transport of particles, but faced with considerable uncertainty in data (such as the sticking coefficient), one of the most useful responses is to use the model to extrapolate in a vicinity in parameter space. For instance, a suitable factorial design of simulations can be used to find out what are the critical parameters in the model and to estimate their values. This will involve using comparison with experiments to determine the values of the missing data.

6.4.2 Analytic Treatment of Neutral Transport in a Trench

We now introduce some analytic approaches to the transport of neutrals in the trench.

A Random Walk in the Trench

Suppose that we have a deep, narrow trench with a mask protecting the edges of the opening above the trench. A neutral etchant enters the trench through the opening in the mask. The etchant has a sticking coefficient γ on the sides of the trench. The probability of bouncing is $(1 - \gamma)$ and the probability of bouncing n_b times is $(1 - \gamma)^{n_b}$. So after n_b bounces the fraction of the neutrals not yet attached to the walls is $(1 - \gamma)^{n_b}$.

The particles diffuse down the trench, bouncing off the walls, until they become attached to a wall. The mean distance they travel down the trench between bounces is comparable to the trench width w. After N steps of a random walk with step size λ the typical distance traveled is about $\sqrt{N}\lambda$. Therefore, in the transport down the trench the particles will have gone about $\sqrt{n_b}w$ in n_b bounces. A typical n_b is $n_b \simeq \gamma^{-1}$, and so the distance particles go down the trench $\sqrt{n_b}w \sim w/\sqrt{\gamma}$.

Strictly, the expected value of n_b is a sum of probabilities times numbers of bounces:

$$\bar{n}_b = 1\gamma + 2\gamma(1 - \gamma) + 3\gamma(1 - \gamma)^2 + \cdots.$$

if we set $n_b = 1$ if the particle stuck the first time it hit. If instead we say it bounced once, and set $n_b = 1$, if the particle bounces the first time it hits and sticks the second time it hits, and set $n_b = 2$ if it sticks the third time it hits, and so on, then

$$\bar{n}_b = 0\gamma + \gamma(1 - \gamma) + 2\gamma(1 - \gamma)^2 + \cdots = \gamma \sum_{n=1}^{\infty} n(1 - \gamma)^n.$$

The number in front of each term in the series is the number of bounces, which is two in the third term we have written out. In that same term, $(1 - \gamma)^2$ is the chance of bouncing twice without sticking, so $\gamma(1 - \gamma)^2$ is the chance of bouncing twice then sticking the next time the particle hits; and so on. Set $\zeta = 1 - \gamma$. Then the sum gives

$$\bar{n}_b = \gamma \sum_{n=1}^{\infty} n\zeta^n = \gamma\zeta \sum_{n=1}^{\infty} n\zeta^{n-1} = \gamma\zeta \frac{d}{d\zeta}\left[\sum_{n=1}^{\infty} \zeta^n\right] = \gamma\zeta \frac{d}{d\zeta}\left[\frac{1}{1 - \zeta}\right].$$

The lower limit in the derivative of the sum can be changed from 1 to 0 because of the derivative which must be found next. Differentiating, we get

$$\bar{n}_b = \gamma\zeta \frac{1}{(1 - \zeta)^2} = \frac{\gamma(1 - \gamma)}{\gamma^2} = \frac{(1 - \gamma)}{\gamma}.$$

Provided $\gamma \ll 1$, this is the same result as above, that is, the typical n_b is γ^{-1}.

If the sticking coefficient γ is 5%, for example, the typical number of bounces is $n_b = 20$; particles will go a mean distance down the trench of about $\sqrt{n_b}w \simeq 4w$. After 64 bounces, $(1 - \gamma)^{64} = (0.95)^{64}$, or about 3% of the particles are left. In 64 bounces the particles will have diffused a mean distance of $8w$. These mean distances do not tell us the shape of the density profile of the diffusing particles, however. To obtain the density profile we shall employ a slightly different formulation of this problem, using a diffusion equation.

Diffusion Equation for Neutrals in the Trench

If the particle speed is v then the diffusion coefficient is $D \simeq \frac{1}{2}v\lambda = \frac{1}{2}vw$. Particles are absorbed at a rate n/τ, where τ is the mean time before a particle is absorbed. Since w/v is the time between bounces, and the typical number of bounces before being absorbed is γ^{-1}, we have $\tau = w/(v\gamma)$.

The continuity equation states that S, the net number of particles produced per second per unit volume, is equal to the sum of two rates. The first is the rate of increase in the number of particles per unit volume, $\frac{\partial n}{\partial t}$. The second is the rate particles leave a unit volume per second, which is $\nabla \cdot \Gamma$ by definition of the divergence, where Γ is the particle flux per unit area. Thus

$$S = \frac{\partial n}{\partial t} + \nabla \cdot \Gamma. \tag{6.23}$$

The particles added to the unit volume in a second (equal in number to S) either go into an increase in the number in the unit volume in one second, which is equal to $\frac{\partial n}{\partial t}$, or they go into the number that leave the unit volume in one second, which is equal to $\nabla \cdot \Gamma$.

In steady state, $\frac{\partial n}{\partial t}$ is zero and so all the particles produced in the unit volume per second, S, go into the number leaving the unit volume per second, $\nabla \cdot \Gamma$.

$$S = \nabla \cdot \Gamma. \tag{6.24}$$

In a purely diffusive process in one Cartesian dimension the flux per unit area per second is $\Gamma = -D\frac{dn}{dx}$. Hence $\nabla \cdot \Gamma = -D\frac{d^2n}{dx^2} = S$ is the appropriate diffusion equation. (For now, x is the depth.)

This steady-state diffusion equation can be written as

$$-D\frac{\partial^2 n}{\partial x^2} = S = -n/\tau \tag{6.25}$$

since the source S is actually the negative of the removal rate n/τ. Using the values of D and τ, we find

$$\frac{1}{2}vw\frac{\partial^2 n}{\partial x^2} = \frac{nv\gamma}{w}, \tag{6.26}$$

and so

$$\frac{\partial^2 n}{\partial x^2} = \frac{2\gamma}{w^2} n = \frac{n}{\ell_d^2}. \tag{6.27}$$

The scale length $\ell_d \equiv w/\sqrt{2\gamma}$ is essentially the same typical distance that particles penetrate into the trench, found above using the first argument, where a distance of $\sqrt{n_b}w \simeq w/\sqrt{\gamma}$ was quoted. The solution of the diffusion equation is

$$n = A \exp(-x/\ell_d) + B \exp(x/\ell_d). \tag{6.28}$$

The boundary condition at the bottom of the trench can be found using ideas from kinetic theory. The first quantity we need to estimate is the number of particles, per second per unit area, striking the bottom of the trench. If the particles were all moving with speed v straight down the trench then a column of particles of height v and unit area would hit each unit area of surface per second. The volume of this column is v and if the neutral density is n, the column contains nv particles. This is the number striking per unit area per second (called the flux per unit area) if the motion is straight toward the surface. Because the flux is isotropic in direction, the number is reduced by a factor of four; the flux per unit area is $\Gamma_{in} = \frac{1}{4} nv$. But of these, only a fraction γ stick and the remaining fraction $(1 - \gamma)$ bounce off. The net flux per unit area, $\Gamma_{net} = \frac{\gamma}{4} nv$, is the difference between the incoming flux per unit area, $\Gamma_{in} = \frac{1}{4} nv$, and the outgoing flux per unit area, $\Gamma_{out} = (1 - \gamma)\frac{1}{4} nv$.

The net flux per unit area in a one-dimensional diffusive process is given by $\Gamma = -D\frac{dn}{dx}$ (where x will be the distance from the top of trench). So at the trench bottom we have

$$\Gamma = \frac{\gamma}{4} nv = -D\frac{dn}{dx}. \tag{6.29}$$

If we use $D \simeq \frac{1}{2} v\lambda$ and the step size $\lambda \sim w$ in our problem, then our boundary condition becomes

$$\frac{\gamma}{4} nv = -\frac{vw}{2} \frac{dn}{dx}, \tag{6.30}$$

or

$$\frac{\gamma}{2w} n = -\frac{dn}{dx}. \tag{6.31}$$

If $\gamma = 0$ there is no sticking and the density gradient is also zero. The diffusion equation yielded an expression for the density $n = Ae^{-x/\ell_d} + Be^{x/\ell_d}$, where $\ell_d = w/\sqrt{2\gamma}$. Using this in the boundary condition, and letting d be the depth of the trench, so that $x = d$ at the bottom, we get

$$\frac{\gamma}{2w}\left\{Ae^{-d/\ell_d} + Be^{d/\ell_d}\right\} = \frac{1}{\ell_d}\left\{Ae^{-d/\ell_d} - Be^{d/\ell_d}\right\}. \tag{6.32}$$

Thus

$$B = Ae^{-2d/\ell_d} \left(1 - \frac{\sqrt{\gamma}}{2\sqrt{2}}\right) \Big/ \left(1 + \frac{\sqrt{\gamma}}{2\sqrt{2}}\right). \qquad (6.33)$$

At $x = 0$, the top of the trench, the density is n_0. The expression for the density, when evaluated at $x = 0$, gives $n(x = 0) = A + B = n_0$, and so $B = n_0 - A$. But $B = fA$, where $f = (2\sqrt{2} - \sqrt{\gamma})/(2\sqrt{2} + \sqrt{\gamma})e^{-2d/\ell_d}$. Therefore $A = n_0/(1 + f)$ and

$$n = \frac{n_0}{1 + f}\left\{e^{-x/\ell_d} + fe^{x/\ell_d}\right\}. \qquad (6.34)$$

Diffusion Equation – Reflecting Walls

We now consider several simpler situations than the case treated so far. If the walls of the trench were not absorbing, but only the bottom absorbed, then the (negative) source term would be zero. The absorption at $x = \ell_d$ is handled through the boundary condition. The diffusion equation is

$$-D\frac{d^2n}{dx^2} = 0, \qquad (6.35)$$

which has the solution $n = Ax + B$. At $x = 0$ the density is $n(x = 0) = n_0 = B$. At the bottom of the trench $x = d$ and $n = Ad + B = Ad + n_0$. The boundary condition we apply, as before, assumes a nonzero γ – but only at the trench bottom:

$$\frac{\gamma}{2w}n = -\frac{dn}{dx}\bigg|_{x=d}. \qquad (6.36)$$

Therefore,

$$\frac{\gamma}{2w}(Ad + n_0) = -A, \qquad (6.37)$$

and

$$A = -n_0/(d + 2w/\gamma). \qquad (6.38)$$

The density is then

$$n = n_0(1 - x/\alpha) \qquad (6.39)$$

with a length scale $\alpha \equiv d + 2w/\gamma$.

This density does not go to zero at the bottom of the trench even when $\gamma = 1$. The expression for the incoming flux $\Gamma_{in} = \frac{1}{4}nv$ that was used does not permit a finite flux when the density is zero. We could attempt to correct the expression for Γ_{in} by evaluating n a mean free path or step size w away from the bottom of the trench, which might be more accurate. If γ is small and w is much smaller than d, these expressions are reasonably accurate. In the limit of a very small trench width w we can assume $w/\gamma \ll d$ provided γ is not too small, in which case $\alpha \simeq d$ and the density is approximately zero at $x = d$. In general the density would extrapolate to zero at a distance $2w/\gamma$ beneath the bottom of the trench. If $\gamma = 0$

this point is at infinity and so the density is completely flat, which guarantees that both expressions for the flux per unit area given below give zero:

$$\Gamma_{\text{net}} = \frac{\gamma}{4} n v \tag{6.40}$$

and

$$\Gamma_{\text{net}} = -D \left. \frac{dn}{dx} \right|_{x=d} = 0. \tag{6.41}$$

These arguments show that if the absorption coefficient of the bottom surface $\gamma = 1$ and the mean free path or the step size is negligible then the density is roughly zero at the bottom surface. If $\gamma = 0$ the gradient of the density is zero at the surface.

Diffusion Equation for a Deep Trench

To describe the density inside a trench whose sidewalls can be etched, we must allow the absorption coefficient γ to be nonzero at the walls as well as the trench bottom. The exponential form of the density found earlier, in Eq. 6.34, is appropriate in this case. The etch rate is presumably proportional to the number of neutrals that stick to unit area of the surface per second, $\Gamma_{\text{in}} = \gamma n v / 4$, and so the etch rate is proportional to the density n. If we assume the trench is deep, so $d \gg \ell_d$, then the factor f is very small and we can approximate the solution found in Eq. 6.34 as

$$n \simeq n_0 e^{-x/\ell_d}. \tag{6.42}$$

Hence the walls etch at a rate that falls exponentially with distance down the trench. When the walls etch enough to change their shape significantly, these estimates for n may not hold. Thus we can only find the initial etch rate in this way.

If the sticking coefficient γ is very small then $\ell_d = w / \sqrt{2\gamma}$ is large. We shall next assume $d / \ell_d \ll 1$. This is the opposite case to the one considered above, and will be shown to give a trivial result. If we do set $d / \ell_d \ll 1$ then $e^{-2d/\ell_d} \simeq 1 - 2d/\ell_d$. Since the first factor in f is $(1 - \sqrt{\gamma}/2\sqrt{2})/(1 + \sqrt{\gamma}/2\sqrt{2}) \simeq 1 - \sqrt{\gamma/2}$ when $\gamma \ll 1$, we have

$$f \simeq (1 - \sqrt{\gamma/2})(1 - 2d/\ell_d) \simeq 1 - \sqrt{\gamma/2} - 2d/\ell_d. \tag{6.43}$$

In these approximations we omitted terms with higher powers of γ than $\sqrt{\gamma}$. We can also omit $\sqrt{\gamma/2}$ compared to $2d/\ell_d = 2d\sqrt{2\gamma}/w$ since $\frac{d}{w} \gg 1$. Finally, since $e^{\pm x/\ell_d} \simeq 1 \pm x/\ell_d$ the density is

$$
\begin{aligned}
n &= \frac{n_0}{2 - 2d/\ell_d} \{1 - x/\ell_d + (1 - 2d/\ell_d)(1 + x/\ell_d)\} \\
&\simeq \frac{n_0}{2} \{(1 + d/\ell_d)(1 - x/\ell_d) + (1 - d/\ell_d)(1 + x/\ell_d)\} \\
&\simeq n_0
\end{aligned}
$$

to the order of approximation we are using. This is not accurate enough to be a useful result. The problem is that if $d/\ell_d \ll 1$, the density is flat.

If we assume γ is very small but d/ℓ_d is not very small then $f \simeq e^{-2d/\ell_d}$ and

$$n = \frac{n_0}{1 + e^{-2d/\ell_d}} \left(e^{-x/\ell_d} + e^{(x-2d)/\ell_d} \right).$$ (6.44)

If we now expand the exponentials $e^{\pm x/\ell_d}$ we find

$$n = \frac{n_0}{1 + e^{-2d/\ell_d}} \left\{ \left[1 - \frac{x}{\ell_d} + \frac{1}{2} \left(\frac{x}{\ell_d} \right)^2 \right] + e^{-2d/\ell_d} \left[1 + \frac{x}{\ell_d} + \frac{1}{2} \left(\frac{x}{\ell_d} \right)^2 \right] \right\}$$

$$= \frac{n_0}{1 + e^{-2d/\ell_d}} \left\{ \left[1 + \frac{1}{2} \left(\frac{x}{\ell_d} \right)^2 \right] \left[1 + e^{-2d/\ell_d} \right] + \frac{x}{\ell_d} \left[e^{-2d/\ell_d} - 1 \right] \right\}$$

$$= n_0 \left\{ 1 + \frac{1}{2} (x/\ell_d)^2 + \frac{x}{\ell_d} \left(\frac{e^{-2d/\ell_d} - 1}{e^{-2d/\ell_d} + 1} \right) \right\}$$

$$\simeq n_0 \left\{ 1 + \frac{x}{2\ell_d^2} (x - 2d) \right\}.$$

At $x = 0$ this is $n(x = 0) = n_0$. At $x = d$ a minimum is reached, with $n(x = d) = n_0\{1 - d^2/2\ell_d^2\}$. The net flux going down the trench is proportional to $-\frac{dn}{dx}$, which is

$$-\frac{dn}{dx} = \frac{n_0}{\ell_d^2} (d - x).$$ (6.45)

This drops to zero at $x = d$ where the slope is zero. The flux is not constant with x but drops linearly with distance owing to losses to the walls.

We shall now rederive the previous approximate expression for the density, in a more straightforward fashion, assuming the density is constant to lowest order. This means the rate of loss to the wall is the same, all the way down the trench. This in turn allows us to obtain a correction to the density.

If the flux going down the trench drops linearly with distance, this implies the loss to the walls is the same for all x, which in turn implies n can be taken to be independent of x in a calculation of the loss rate to the wall. The rate of loss to the wall is n/τ with $\tau = w/\gamma v$.

If we approximate this loss to the wall as n_0/τ then

$$-D\frac{d^2n}{dx^2} = -\frac{1}{2} vw \frac{d^2n}{dx^2} = -n/\tau \simeq -\frac{n_0 \gamma v}{w}$$ (6.46)

and therefore

$$\frac{d^2n}{dx^2} = \frac{2\gamma}{w^2} n_0 = \frac{n_0}{\ell_d^2}.$$ (6.47)

The only difference from the first equation we had is that n_0 appears on the right-hand side. Integrating, we get

$$\frac{dn}{dx} = A + \frac{n_0 x}{\ell_d^2}$$ (6.48)

and

$$n = B + Ax + \frac{n_0 x^2}{2\ell_d^2}. \tag{6.49}$$

The boundary conditions are $n(x = 0) = n_0 = B$ and

$$\Gamma_{x=d} = \frac{\gamma}{4}nv = -D\frac{dn}{dx} = -\frac{1}{2}vw\frac{dn}{dx}. \tag{6.50}$$

Since this equation implies that $\frac{dn}{dx} = -\frac{\gamma n}{2w}$ and $\gamma \ll 1$ this boundary condition also implies $\frac{dn}{dx}$ is very small at $x = d$. Using $\frac{dn}{dx} = A + n_0 x/\ell_d^2$ and setting $x = d$ gives

$$A + \frac{n_o d}{\ell_d^2} \simeq 0. \tag{6.51}$$

So our solution for n is

$$n = n_0 - \frac{n_0 dx}{\ell_d^2} + \frac{n_0 x^2}{2\ell_d^2}, \tag{6.52}$$

which is the same answer we obtained before, derived more directly.

Exercises

1. When metal atoms ("adatoms") are deposited on a crystal surface, they often diffuse until they are trapped by a defect on the surface. Defects may be line or point defects. Suppose there are N_d point defects per square meter of a flat crystal surface. Ignore line defects. Each point defect has a radius a_d, such that $N_d \pi a_d^2 \ll 1$. What is the significance of $N_d \pi a_d^2 \ll 1$? The adatoms move randomly on the crystal surface, taking steps of size $\Delta \sim a_c$, until captured by a defect. Here a_c is the spacing of atoms in the crystal. If the adatoms moved in a straight line, about how far, λ_{ad}, would they travel on average before hitting a defect? Since the adatoms diffuse on the surface, are they more or less likely to hit a defect before traveling a net distance λ_{ad}? Explain why.

    Write a Monte Carlo program to simulate this diffusion. Place defects at random positions on the surface, with density N_d. Launch adatoms at random positions on the surface. Follow adatoms until they strike defects. The region should be periodic, that is, if adatoms leave one edge of the simulation, they should be replaced at the opposite edge. Record the net distance traveled in each case, and plot the results as a histogram. Use $a_d = a_c$ equal to the lattice spacing for silicon, and choose N_d to correspond to one defect in each of 10^2, 10^4, and 10^6 silicon atoms on the surface.

2. This problem illustrates the use of transition probabilities in describing particle transport. It uses an analogy, which we might call the "drunken soccer fans traveling by train" problem. Suppose we have $N_f = 1,000$ soccer fans, who are attempting to travel by train, from the third station on a railway line with ten stations, to the eighth station. Whenever they board a train, they have a

0.5 chance of going in the right direction (except at the ends of the line). The trains stop at every station, and the fans stay on the train with probability 0.8 whenever the train stops in a station. The trains are synchronized, so that each station has one train leaving in each possible direction, at each hour interval.

What are the possible choices of dependent variable? For example, one could calculate the number of fans changing trains at each station, after an integer number of train stops at successive stations. This choice would require knowledge of the probability that a fan goes from any given station, to any other given station, without disembarking from the train on the way. Alternatively, one might use the number of fans changing at each station, as well as the number of fans on trains going up the line who did not change and the number on trains going down the line who did not change. With these dependent variables calculated each time the trains stop, the probabilities needed are related only to movements between one station and the next. What are the advantages and disadvantages of each? Set the problem up using one of these formulations, and solve it using a computer; that is, find where the fans are after each hour, when the trains all stop in stations. Try the calculation, assuming the fans stop in the eighth station, as soon as they arrive there, and do not leave. Repeat, assuming that they do not stop at their destination but keep on traveling.

3. Write a Monte Carlo (MC) program to simulate particle motion inside a trench. The walls absorb particles with probability γ. Particles leaving the wall are distributed isotropically. Gas phase collisions are negligible. If $\gamma = 0.1$, the trench has an aspect ratio of 10 and is infinite in the z direction, and particles enter the top of the trench isotropically, use the MC program to calculate the density n. Compare the result to the analytic approximation found above. Repeat the MC calculation, if the vertical axis of the trench is the x direction, its horizontal axis is the y direction, and the incoming particles all enter traveling in the direction $-\alpha\hat{x} - \sqrt{1 - \alpha^2}\hat{y}$ for various values of α such that $\alpha < 1$.

4. Evaluate Eq. 6.34 and plot n as a function of x, for $\gamma = 0.05$ and $d = 2w, 5w$ and $10w$. Repeat for $\gamma = 0.01$.

7

Evolution of the Trench

7.1 Evolution of the Trench

In this chapter we examine a number of physical processes that affect the evolution of the shape of a trench. Etching by neutrals alone is considered first, followed by charged particle effects and behavior. This leads on to charging of the trench walls and the effect of the charging on the charged particles.

The calculation of surface processes from first principles is not feasible at present. Atomistic studies of surface behavior have been widely used, employing accurate interatomic potentials [13,12,11]. Recent attempts to calculate correlation energies, with a view to improving methods such as Density Functional Theory (DFT), are described in Refs. [166]–[168]. For now, however, we stick to phenomenological approaches.

7.1.1 Etching by Neutrals

If the distribution of neutrals is isotropic and uniform and neutrals are responsible for the etching, then each section of the surface is etched at the same rate whatever the position or orientation of the section. If the surface is divided into sections that are all horizontal or vertical then each corner moves at $45°$ to the vertical. Suppose the distribution is isotropic, and that the density of neutrals varies slowly enough in space so that the sections on each side of any corner etch at roughly the same rate. This means the $45°$-rule still holds approximately for each corner even though different corners may move at different rates. The best way to see how this works is to consider a few examples.

The first example we should consider consists of the etching with an isotropic, uniform distribution of neutrals. The surface shifts outward but retains the same shape (Fig. 7.1). Vertical strings of sections stay vertical and horizontal strings of sections stay horizontal, even though the corners of the sections move at $45°$.

The next example will be etching with an isotropic neutral distribution, but with a neutral density that is nearly zero at the bottom of the trench.

As long as the etch rate of successive sections is nearly the same, the corners still move at nearly $45°$. The etch rate goes to zero at the bottom, however, so the etch rate changes by a large fraction on going to the last section. Any corner at the end of a horizontal surface that has zero etch rate can only move horizontally.

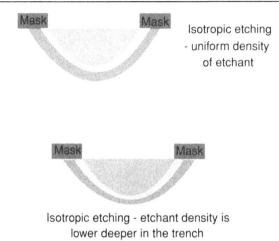

Fig. 7.1. Etching with isotropic, uniform density of neutrals and with an isotropic, nonuniform density.

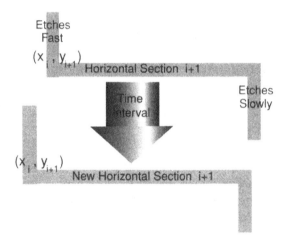

Fig. 7.2. Evolution of a horizontal section. In this figure x_i is the horizontal coordinate of vertical section i and y_{i+1} is the vertical coordinate of horizontal section $i+1$.

Any corner of a vertical surface, with no etching of the vertical surface, can only move vertically; in this case the corner goes straight down. We shall examine the shape evolution, by considering a surface made up of horizontal and vertical (HV) sections and how the HV sections move. A surface made up of HV sections is sketched in Fig. 7.2. (We could have handled the variation in etch rate by moving the corners at say 45° and allowing the orientation of the sections to vary, but this is less convenient.)

The next two sections give simple analytic estimates of how the orientation of the surface evolves.

Shape Evolution I

Consider a surface such as is shown in Figs. 6.3 and 7.2. Suppose vertical section one moves a distance Δ_1 and horizontal section two moves Δ_2. This would correspond to the surfaces next to the words "old surface" in Fig. 6.3. Δ_1 is the distance dx; Δ_2 is the distance dy. Then the interior corner between sections one and two moves Δ_1 horizontally and Δ_2 vertically and so it moves at an angle of θ_1 to the vertical, where $\tan \theta_1 = \Delta_1/\Delta_2$. Similarly, if θ_2 is the angle the motion of the next (exterior) corner makes to the vertical, $\tan \theta_2 = \Delta_3/\Delta_2$.

Suppose that the neutral density depends only on the depth, which in this section we denote by x. (On other occasions we use coordinates (x, y), with x the horizontal coordinate and y the vertical coordinate, but when x is used alone it denotes the depth.) Thus the etch rate only depends on depth. Then Δ_i of the ith section is approximately $\Delta_i = \Delta(x_i) = \eta \, n(x_i)$, where x_i is the depth of the midpoint of the ith section and η is a constant proportional to the time step δt; that is, $\eta = r \delta t$.

We expect $n(x)$ to decrease with depth. This means that at the interior corner after vertical section i and before horizontal section $(i + 1)$, the angle at which the corner moves, θ_i^{int}, is given by

$$\tan \theta_i^{\text{int}} = \frac{\Delta_i}{\Delta_{i+1}} = \frac{n(x_i)}{n(x_{i+1})}. \tag{7.1}$$

If the vertical ith section has its midpoint at x_i and is of length δx_i then the depth of the horizontal $(i + 1)$th section is $x_{i+1} = x_i + 0.5 \, \delta x_i$. Expanding $n(x_{i+1})$ in a Taylor series gives

$$n(x_{i+1}) \simeq n(x_i) + \frac{dn}{dx} \frac{\delta x_i}{2} \tag{7.2}$$

and

$$\tan \theta_i^{\text{int}} = \left[1 + \frac{\delta x_i}{2n} \frac{dn}{dx} \right]^{-1} \simeq 1 - \frac{\delta x_i}{2n} \frac{dn}{dx}. \tag{7.3}$$

Now suppose θ_i^{int} is close to $\pi/4$, so that $\theta_i^{\text{int}} = \frac{\pi}{4} + \delta\theta_i^{\text{int}}$. Now $\tan(\pi/4 + \delta\theta) \simeq 1 + 2\delta\theta$ if $\delta\theta$ is small, so in this case

$$\delta\theta_i^{\text{int}} \simeq -\frac{\delta x_i}{4n} \frac{dn}{dx}, \tag{7.4}$$

which is positive since $\frac{dn}{dx}$ is negative. This is the amount by which the angle to the vertical made by the interior corner's movement exceeds $\pi/4$.

At the exterior corner between horizontal section j and vertical section $j + 1$, the angle to the vertical made by the corner's motion is θ_j^{ext}, where

$$\tan \theta_j^{\text{ext}} = n(x_{j+1})/n(x_j) \simeq \left[1 + \frac{\delta x_{j+1}}{2n} \frac{dn}{dx} \right]. \tag{7.5}$$

The exterior corner moves at an angle to the vertical of $\pi/4 + \delta\theta_j^{\text{ext}}$, where

$$\delta\theta_j^{\text{ext}} \simeq \frac{\delta x_{j+1}}{4n}\frac{dn}{dx}. \tag{7.6}$$

If we define the local length scale for the density to be L_d, where

$$\frac{1}{L_d} \equiv -\frac{1}{n}\frac{dn}{dx}, \tag{7.7}$$

then

$$\delta\theta_i^{\text{int}} = \frac{\delta x_i}{4L_d} \tag{7.8}$$

and

$$\delta\theta_j^{\text{ext}} = -\frac{\delta x_{j+1}}{4L_d}. \tag{7.9}$$

These confirm that if the neutral density decreases with depth, the horizontal sections grow at the expense of the vertical sections. If the density decays exponentially as we found above for a deep trench, $n \simeq n_0\exp(-x/\ell_d)$, then the length scale L_d, which is given by $L_d^{-1} = -\frac{1}{n}\frac{dn}{dx}$ is just $L_d = \ell_d = w/\sqrt{2\gamma}$.

Shape Evolution II

Now suppose we have a string of sections, and the first part of each vertical section is α times as long as the second part of the horizontal section that came before it. Similarly the second part of each vertical section is α times as long as the first part of the horizontal section that follows it. The string is overall at an angle θ_s to the vertical, where $\tan\theta_s = \alpha^{-1}$. The angle θ_s will change during etching. The section of surface is shown in Fig. 7.3.

Section 1 moves down by a distance $\Delta_1 = \eta n(x_1)$. Section 2 moves left by $\Delta_2 = \eta n(x_2)$ (if we evaluate the etch rate at the position x_2, which is shown on the figure) and $x_2 = x_1 + \alpha\ell_1^2$. So $n(x_2) = n(x_1) + \alpha\ell_1^2\frac{dn}{dx}$ and $\Delta_2 = \eta(n(x_1) + \alpha\ell_1^2\frac{dn}{dx})$.

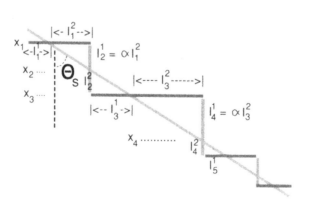

Fig. 7.3. String of sections at an angle θ_s to the vertical.

If we define $\beta = \eta\alpha\frac{dn}{dx}$ then

$$\Delta_2 = \eta n(x_1) + \beta\ell_1^2. \tag{7.10}$$

The center of Section 3 is at $x_3 = x_1 + (\ell_2^1 + \ell_2^2) = x_1 + \alpha(\ell_1^2 + \ell_3^1)$ and

$$\Delta_3 = \eta n(x_1) + \beta(\ell_1^2 + \ell_3^1). \tag{7.11}$$

Finally, the center of Section 4 is at $x_4 = x_1 + \alpha(\ell_1^2 + \ell_3^1 + \ell_3^2)$ and

$$\Delta_4 = \eta n(x_1) + \beta(\ell_1^2 + \ell_3^1 + \ell_3^2). \tag{7.12}$$

ℓ_2', the new length of Section 2 is the old length ℓ_2 plus the distance Section 3 moves, minus the distance Section 1 moves.

$$\ell_2' = \ell_2 + \Delta_3 - \Delta_1 = \ell_2 + \beta(\ell_1^2 + \ell_3^1). \tag{7.13}$$

ℓ_3', the new length of Section 3 is similarly

$$\ell_3' = \ell_3 + \Delta_2 - \Delta_4 = \ell_3 - \beta(\ell_3^1 + \ell_3^2) = \ell_3(1 - \beta). \tag{7.14}$$

By analogy with ℓ_2', the new length of Section 4 is

$$\ell_4' = \ell_4 + \beta(\ell_3^2 + \ell_5^1). \tag{7.15}$$

If the horizontal sections are uniform then $\ell_1^1 = \ell_1^2 = \ell_3^1 = $ etc. Then $\ell_2' = \alpha\ell_1 + \beta\ell_1$ and since $\ell_3' = \ell_3(1 - \beta)$ in any case we can find the next angle θ_s' from the new ratio α':

$$\alpha' = \ell_2'/\ell_3' = (\alpha\ell_1 + \beta\ell_1)/\ell_3(1 - \beta) = (\alpha + \beta)/(1 - \beta) \tag{7.16}$$

since, for uniform sections, $\ell_3 = \ell_1$.

The new angle θ_s' is given by

$$\tan\theta_s' = (\alpha')^{-1} = \frac{1 - \beta}{\alpha + \beta}. \tag{7.17}$$

The difference between θ_s' and θ_s is $\delta\theta_s$. Therefore

$$\tan(\theta_s + \delta\theta_s) = \frac{1 - \beta}{\alpha + \beta}. \tag{7.18}$$

For small $\delta\theta_s$ and small β this becomes

$$\tan\theta_s + \frac{\partial}{\partial\theta}(\tan\theta_s)\delta\theta_s \simeq \frac{1}{\alpha}\left(\frac{1 - \beta}{1 + \beta/\alpha}\right) \simeq \frac{1}{\alpha}(1 - \beta - \beta/\alpha). \tag{7.19}$$

So since $\tan\theta_s = 1/\alpha$,

$$\frac{\delta\theta_s}{\cos^2\theta_s} = -\frac{\beta}{\alpha}(1 + 1/\alpha). \tag{7.20}$$

Since $\beta = r\alpha \frac{dn}{dx}\delta t$ and $\cos^2 \theta_s = \alpha^2/(1+\alpha^2)$, we have

$$\frac{\delta\theta_s}{\delta t} \simeq \frac{d\theta_s}{dt} = -r\alpha\frac{dn}{dx}\frac{\cos^2\theta_s}{\alpha}(1+1/\alpha) = -r\frac{dn}{dx}\frac{(1+\alpha)\alpha}{(1+\alpha^2)}. \qquad (7.21)$$

Since $\frac{dn}{dx}$ is negative, this shows that the angle to the vertical increases with time, at this rate.

7.2 Computation of the Surface Shape

The surface of the cross section of the trench has been described as a series of straight sections. These sections are labeled by an integer i. Since the sections are alternately horizontal then vertical, we can say that sections with even values of i are horizontal (or vice versa). One advantage of this method of describing the surface is that it makes it possible to describe how the cross section evolves, in a straightforward way.

To specify the shape of the surface, all that we need to know is where each section is, with respect to one coordinate axis. The horizontal sections must have their height, y_i, specified. The vertical sections must have their horizontal position, x_i, specified. From these we can deduce where the sections meet. For instance, vertical section i is at x_i; the next section is horizontal section $i+1$, which is at y_{i+1}. They meet at (x_i, y_{i+1}). Note that we were using x to mean the depth earlier. When the two coordinates, (x, y), are both used, then x is the horizontal coordinate and y is the vertical coordinate.

There are some circumstances where the shape of the surface could become complicated enough to cause problems, which we shall not go into here.

To calculate the evolution of the surface, we need to know the etch rate of each of the sections. If the etching is purely chemical, and the trench has a high aspect ratio, then the density of etchants, and hence the etch rate, might depend only on the depth in the trench, y. Then knowing the etch rate $r(y)$ will allow us to set up a numerical calculation of the surface evolution.

7.3 The Evolution of a Trench with Ion-Assisted Chemical Etching

The addition of ions, which assist the etching, will tend to increase the rate of etching of horizontal sections since ions tend to come into the trench moving predominantly downward. The flux of ions will probably vary less with depth in the trench than that of the neutrals, again because the ions are expected to move almost straight down the trench. This directed ion motion should mean collisions with the walls do not occur frequently. Thus many of the ions reach the bottom of the trench. We shall return to this issue, of where the ions actually go, shortly.

To model the etching we must choose a form for the etch rate in the presence of ions and neutral radicals. Most etchers are designed so that both ions and neutrals are needed for etching, but one of the species may only be needed up to a threshold beyond which adding more will have no effect. Alternatively, the rate of etching may be determined by whichever of the species needed for etching is in shortest supply.

An example of a situation where the flux exhibits a threshold for etching would be in a situation where ions were needed to clear a depositing film off the surface. As long as the surface was clear of film, the underlying surface can be etched chemically. The ion flux, or the ion energy flux, has to be above some value to keep the film from growing. If the ion flux were lower than the threshold, then the film would grow, although more slowly than if there was no ion flux. If the ion flux higher than the threshold, again no film would form. Thus the rate of etching depends on whether the ion flux is above or below threshold; it does not depend on how far the flux is away from threshold. This picture is oversimplified in several ways, however. First, the ion energy flux is usually the most important quantity, so only if the ion energy is roughly constant will the ion particle flux determine the threshold, above which etching can take place directly. The ion energy is approximately equal to the energy the ions acquire in crossing the sheath, provided they have no collisions while crossing the sheath and provided the potential inside the trench is negligible. This ambiguity obscures what the threshold is. Second, the ion energy flux needed to keep the surface clear will probably depend on the deposition rate of the radicals, which are needed for the film to grow.

In other cases the ions and neutral radicals both appear to contribute directly to the etching. If ions and neutrals are required to participate in a reaction, then they are required in the right proportions. The excess of one species, if there is an excess, goes unused because when all of the other species are used up, the reaction stops. In the ECR etcher we use as an extended example of CF_4 plasma chemistry, there frequently appears to be an excess of the neutral radical etchant, fluorine. The fluorine radical may attach to the surface and the surface may, in the presence of enough F, be able to saturate so that no more F can attach. The saturation would limit the available F for etching, but if enough F is available to meet the needs for etching at the rate the ions can sustain, the etch rate would be controlled by the ion (energy) flux.

7.3.1 Ion Motion in the Sheath

We shall next suppose that there is enough of the neutral radical so that the ion energy flux is the rate determining factor for the etching. The trench profile evolution will then depend on the angular distribution of ions in the trench, which determines the flux of ions to each portion of the surface. If there is little or no electric field inside the trench then the distribution of ions is determined by what happens before they enter the trench.

In summary, the etch rate will at first be assumed to be proportional to the ion energy flux, but the ion kinetic energy will be taken to be approximately the energy picked up in crossing the sheath, qV_{sh}. Thus, the etch rate will be proportional to the ion flux provided V_{sh} is constant and the mean free path $\lambda \gg t_s$, the sheath thickness: In reality the etch rate might depend on the ion flux in a nonlinear fashion but at least up to some level set by the amount of neutral etchant available, we can expect the etch rate to be roughly proportional to the ion energy flux:

$$R^{\text{etch}} = r N_i v_{i\perp} q V_{sh} = r' N_i v_{i\perp}. \qquad (7.22)$$

Here N_i is the ion density, $v_{i\perp}$ is the ion velocity normal to the surface being etched, and r is a rate constant. The quantity $r' = rq V_{sh}$ is assumed to be constant. All of the ion's energy is assumed to be given to whatever surface it strikes, and so the energy appearing in the energy flux is qV_{sh}, independent of which surface is involved.

If the ions fall through the sheath potential V_{sh} with no scattering, their velocity normal to the wafer surface is roughly $v_\perp = \sqrt{2qV_{sh}/m_i}$, where m_i is the ion mass. In the direction parallel to the wafer surface, their velocities are expected to have a thermal distribution, because that was the distribution they had in parallel velocity $v_\parallel$ before they fell through the sheath. The sheath electric field could have a small component of electric field $E_\parallel$ parallel to the surface but we ignore this possibility for now. If $E_\parallel = 0$ in the sheath then $v_\parallel$ does not change when the ion crosses the sheath.

The typical speed in a thermal distribution with temperature T_i is (about) $v_{th} = \sqrt{2k_BT_i/m_i}$, where k_B is Boltzmann's constant. Since $v_\parallel$ is comparable to v_{th} whereas $v_\perp \simeq \sqrt{2qV_{sh}/m_i}$, the ratio of these velocities is

$$\frac{v_\parallel}{v_\perp} = \sqrt{\frac{k_BT_i}{qV_{sh}}}. \tag{7.23}$$

Now qV_{sh} may be of the order a few times k_BT_e or higher (the sheath has to repel most of the electrons); the applied voltage may force V_{sh} to be larger, and so

$$\frac{v_\parallel}{v_\perp} \lesssim \sqrt{\frac{T_i}{T_e}}. \tag{7.24}$$

If T_i is the neutral gas temperature, and since setting T_i to be about room temperature would be an underestimate, then $k_BT_i \sim \frac{1}{30}$ eV whereas $k_BT_e \sim 2$ eV, for example, and

$$\frac{v_\parallel}{v_\perp} \simeq \sqrt{\frac{1}{60}} \sim \frac{1}{8}. \tag{7.25}$$

This shows that unless the sheath voltage is increased, the motion is significantly affected by the thermal contribution to $v_\parallel$. A particle with this ratio of velocities moves at an angle θ to the vertical given roughly by $\tan\theta = v_\parallel/v_\perp$. For small θ, $\tan\theta \sim \theta$, provided θ is in radians; so $\theta \simeq \frac{1}{8}$ radians $\simeq 8°$.

The horizontal flux $\Gamma_\parallel$ striking the side of the trench is proportional to $v_\parallel$ whereas the vertical flux $\Gamma_\perp$ striking the bottom is proportional to $v_\perp$. The thermal energy associated with $v_\parallel$ is small but $v_\parallel$ itself is significant and apparently is enough to make the horizontal etch rate more than 10% of the vertical etch rate.

Effects of Ion–Neutral Collisions

Elastic collisions of ions with neutrals tend to increase the energy associated with ion motion parallel to the surface. If there were no ion–neutral collisions, the ion motion would be highly directed, being nearly straight toward the surface, with $v_\perp$ being large and $v_\parallel$ relatively small. Elastic collisions decrease the ion energy and tend to randomize the velocity, turning part of $v_\perp$ into $v_\parallel$. Since $v_\parallel$ was initially small it is very unlikely that $v_\parallel$ will give up energy to $v_\perp$ as the result of a collision.

There are two types of ion collisions with neutrals that are likely to be important: elastic collisions and charge-exchange collisions. The importance of charge-exchange collisions will depend on the gas mixture. If charge exchange does occur, the ion is neutralized, producing a fast neutral with roughly the velocity the ion had before the collision.

This fast neutral is moving rapidly out of the plasma and may make a contribution to etching similar to that of a fast ion. The new ion produced in the charge exchange is expected to have roughly the velocity of the original neutral, and so its $v_{\parallel}$ and $v_{\perp}$ are thermally distributed at the neutral temperature. After the collision $v_{\perp}$ changes – the particle accelerates in the electric field. However, because $v_{\parallel}$ changes little in the sheath electric field, this ion will arrive at the surface with a thermal $v_{\parallel}$ and a $v_{\perp}$ that is high, but not as high as if the ion fell all the way through the sheath without a collision. In other words, the charge-exchange collision by itself would have little effect on the distribution of $v_{\parallel}$, if $v_{\parallel}$ were already thermally distributed. The collision does decrease somewhat the $v_{\perp}$ of the ion when it reaches the surface.

It could be that $v_{\parallel}$ was very large before the charge exchange, perhaps due to an earlier elastic ion–neutral collision. In that case the charge exchange will tend to decrease $v_{\parallel}$. The subsequent increase of $v_{\perp}$ in the sheath field will leave the ion moving nearly straight toward the surface. It can thus be argued that charge-exchange ion collisions might actually decrease the spreading of the ion distribution caused by elastic collisions, since elastic collisions in the sheath typically lead to higher $v_{\parallel}$ than the thermal distribution of $v_{\parallel}$ would give. The next few paragraphs illustrate this point in more detail.

Suppose an ion that is crossing the sheath has an elastic collision and that on average the ion velocity after the elastic collision is isotropically distributed. (The collision is not likely to be isotropic in reality. The role of anisotropic scattering is discussed in Ref. [82].) Then immediately after the collision the ion has the same chance of going in any direction. If we imagine a sphere centered on the point where the collision occurred, the particle has equal chances of coming out of the sphere through any equal areas on the surface of the sphere. The direction normal to the wafer surface defines an axis within the sphere, from which we can measure an angle θ. The area within a range $\Delta\theta$ of $\theta = 0$ going directly away from the surface or within $\Delta\theta$ of $\theta = \pi$ going straight toward the surface is (in each case) $\pi(r\,\Delta\theta)^2$, with r being the sphere's radius, this area subtending a solid angle of $\pi(\Delta\theta)^2$. The area within $\Delta\theta$ at $\theta = \pi/2$ (going parallel to the surface) is $2\pi r(r\,\Delta\theta)$ and has solid angle $2\pi\,\Delta\theta$. In other words, the chance of going in a direction close to $\theta = 0$ is small, for isotropic scattering.

If the scattered ion is moving away from the surface immediately after the scatter and has no other collisions first, then it will continue to move away until the electric field toward the surface can turn it around and bring it back. When it again reaches the same distance from the surface as it was when it scattered, it will have reversed $v_{\perp}$, its velocity normal to the surface. It may have moved sideways parallel to the surface in the elapsed time. In most cases identical ions with the same velocity will move sideways and take each others' places; thus the sideways movement has little effect (except near the outside edge of the wafer, where there may not be ions on the outside of the wafer to change places with others further in). The effect of the electric field in pushing ions back toward the surface means that we can assume that the ions all scatter isotropically, but within the half of the sphere corresponding to their having a $v_{\perp}$ that is toward the surface.

Suppose the ion fell from the sheath edge, a height x_s above the wafer. Its chance of colliding in the range of distances dx after falling a distance x is

$$p(x)dx = e^{-x/\lambda_i} dx/\lambda_i \tag{7.26}$$

where x is now being measured downward from the top of the sheath, and λ_i is assumed to be constant, which is not a very accurate assumption. (Integrating this expression from $x = 0$ to $x = \infty$ gives the total chance of a collision in going an infinite distance: $\int_0^\infty p(x)dx = 1/\lambda_i \int_0^\infty e^{-x/\lambda_i} dx = -e^{-x/\lambda_i}|_0^\infty = 1$, which shows that the normalization of $p(x)$ is correct.) If $\lambda_i \gg x_s$ then $e^{-x/\lambda_i} \simeq 1$ and $p(x) = \frac{1}{\lambda_i}$. Thus the ion has a nearly constant, small chance of colliding between x and $x + dx$ given by $p(x)dx$, which is nearly independent of x. Only a fraction of about $x_s/\lambda_i \ll 1$ of the ions collide at all.

If the electric field in the sheath is $E(x)$ and is in the x direction, then the ion picks up kinetic energy $T_i = \int_0^{x_c} qE(x)dx$ before it has a collision at $x = x_c$. If the electric field varies linearly in the sheath, corresponding to constant charge density in the sheath, then $E(x) = E'x$, since $E(x=0) = 0$ at the plasma edge of the sheath. Now $E' \equiv \frac{dE}{dx} = \rho/\varepsilon_o = qn_i/\varepsilon_o$, according to Poisson's equation, assuming negligible numbers of electrons in the sheath. Then by the time it reaches a point a distance x into the sheath, the ion has kinetic energy $T_i = \int_0^x qE'xdx = qE'x^2/2$. The ion speed v_i is given by $T_i = \frac{1}{2}mv_i^2 = \frac{1}{2}qE'x^2$ and so

$$v_i = (qE'/m)^{\frac{1}{2}}x. \tag{7.27}$$

The collision takes place at $x = x_c$. Before the collision the ion velocity was pointing roughly along $\theta = \pi$. After the collision it will be moving at some angle θ_c to the outward normal, and a fraction of the ion energy is lost, so immediately after the collision

$$v_\perp = -\alpha(qE'/m)^{1/2}x_c \cos\theta_c \tag{7.28}$$

and

$$v_\parallel = \alpha(qE'/m)^{1/2}x_c \sin\theta_c, \tag{7.29}$$

where $(1 - \alpha^2)$ is the fraction of the ion energy lost in the collision and the kinetic energy associated with motion toward the surface is $T_\perp = \frac{1}{2}mv_\perp^2 = \frac{1}{2}\alpha^2 qE'x_c^2 \cos^2\theta_c$. Similarly, the parallel kinetic energy is $T_\parallel = \frac{1}{2}\alpha^2 qE'x_c^2 \sin^2\theta_c$.

The energy $T_\parallel$ does not change in the electric field as the ion continues to fall but $T_\perp$ does. If there are no more collisions, then by the time the ion reaches the bottom of the sheath after falling a total distance of x_s, the kinetic energy $T_\perp$ associated with $v_\perp$ increases by an amount $\delta T_\perp$ such that

$$\delta T_\perp = \int_{x_c}^{x_s} qE'xdx = \frac{1}{2}qE'(x_s^2 - x_c^2). \tag{7.30}$$

The final energy of the ion T^{final} is

$$T^{\text{final}} = \frac{1}{2}qE'(x_s^2 - (1-\alpha^2)x_c^2) \tag{7.31}$$

since $(1 - \alpha^2)$ is the fraction of the ion energy lost at position x_c. The kinetic energy associated with ion motion parallel to the surface, divided by this total final kinetic energy, gives $v_\parallel^2/(v_\parallel^2 + v_\perp^2)$, which equals $\sin^2\theta$:

$$\frac{T_\parallel}{T^{\text{final}}} = \alpha^2 x_c^2 \sin^2\theta_c/\left(x_s^2 - (1 - \alpha^2)x_c^2\right) = \sin^2\theta, \tag{7.32}$$

where θ is the angle the ion motion makes with the outward normal when it hits the wafer.

If $(1 - \alpha^2)x_c^2$ is small compared to x_s^2 then $\sin^2\theta \simeq \alpha^2 x_c^2 \sin^2\theta_c/x_s^2$ or $\sin\theta \simeq \pm\alpha\frac{x_c}{x_s}\sin\theta_c$. If in addition $\alpha x_c/x_s \ll 1$ then θ is close to π, $\theta = \pi - \delta\theta$, and $\delta\theta \simeq \frac{\alpha x_c}{x_s}\sin\theta_c$. To proceed further we need to know the distribution of $(\alpha \sin\theta_c)$, which since α and θ_c depend on each other should be treated as one quantity, and of x_c/x_s. The latter is straightforward as long as $x_s/\lambda_i \ll 1$; all values of x_c/x_s from 0 to 1 are equally likely. To find $\alpha \sin\theta_c$ we need to consider the collision dynamics. The value of α for a given θ_c can be found from conservation of energy and momentum, but the probability of a particular θ_c depends on details of the cross section, which are only known approximately.

The implication of this analysis is that for $\lambda_i \gg x_s$, only a small fraction equal to x_s/λ of the ions collide while crossing the sheath. The remaining fraction $(1 - x_s/\lambda)$ have a thermal spread in $v_\parallel$, whereas $v_\perp$ for these particles is given roughly by $\frac{1}{2}mv_\perp^2 = qV_{\text{sh}}$. The ions that have elastic collisions with neutrals lose some energy, depending on the details of the collision and on where the collision happened. They arrive at an angle to the surface $\delta\theta \sim \alpha\frac{x_c}{x_s}\sin\theta_c$, where $\alpha\sin\theta_c \lesssim 1$ and $0 < x_c/x_s \le 1$. Charge-exchange collisions leave the ions with $v_\parallel \sim v_{\text{th}}$, the neutral thermal velocity, and a similar value for $v_\perp$. Thus charge-exchange collisions result in a smaller $v_\parallel$ than elastic collisions do. The total mean free path λ_i is related to the elastic mfp, λ_i^{el}, and the charge-exchange mfp, λ_i^{cx}, by

$$\frac{1}{\lambda_i} = \frac{1}{\lambda_i^{\text{el}}} + \frac{1}{\lambda_i^{\text{cx}}}. \tag{7.33}$$

The chance of a collision while crossing the sheath is x_s/λ_i, (provided that chance is small) which is the sum of the chance of having an elastic collision, $x_s/\lambda_i^{\text{el}}$, and the chance of having a charge-exchange collision, $x_s/\lambda_i^{\text{cx}}$, as the above equation shows when multiplied by x_s.

7.4 Charging of the Trench Walls

Once the ions enter the trench they have only a small distance to go before they hit the trench bottom or sidewalls. The trench dimension t_d is very small compared to the mean free path λ_i, making collisions very unlikely. Furthermore, t_d is also very small compared to the Debye length λ_D, which is the scale on which the charge density per unit volume can change the electric field. Put another way, for the given charge density per unit volume, that charge will have little effect in a region as small as the trench. To show this, we can find the charge inside the volume of the trench. This charge turns out to be much too small to affect the electric field much. To find the change in E within the trench, we begin with Poisson's equation written in the form

$$\frac{dE}{dx} = \frac{\rho}{\varepsilon},$$ (7.34)

which is a one-dimensional version of the equation. Integrating this equation downward in x from the top toward the bottom of the trench, we find $\Delta E = \int_o^{t_d} \frac{\rho}{\varepsilon} dx$. If ρ is the ion charge density, there being little electron charge density, then $\rho \simeq n_i e$. The permittivity $\varepsilon = \varepsilon_o = 8.8 \times 10^{-12}$ F/m, and $n_i \lesssim 10^{17}$ m^{-3}, and so $n_i e \lesssim 1.6 \times 10^{-2}$ and $\rho/\varepsilon \lesssim 1.6 \times 10^9$ V/m. This means that if $t_d \sim 1\,\mu$m $= 10^{-6}$ m, then $\Delta E \lesssim 1.6 \times 10^3$ V/m.

This appears to be a large field, but the electrostatic potential change in the trench due to this field can be found using $E = -\frac{dV}{dx}$ and is $\Delta V \sim -(\Delta E)t_d$. Then $\Delta V \sim 10^{-3}$ volts, which is quite negligible.

Another possible source of electric fields in the trench is any charge that builds up on the surface, for instance if part of the surface is an insulator and the fluxes per unit area of ions (Γ_i) and electrons (Γ_e) are not initially equal. If the fluxes are not equal and if the charge cannot leak away, then the surface charge will build up, until the field due to the surface charge can exert enough force to make $\Gamma_e = \Gamma_i$. When the fluxes are equal the potentials must be of order T_e (expressed in eV) to deflect the charged particles. Both electrons and ions have typical energies of T_e or more by the time they enter the trench.

To find the electric field in the trench it is only necessary to solve Laplace's equation, which is Poisson's equation with ρ set to zero, because we showed that ρ has a negligible effect in the trench. The boundary conditions at the trench walls bring in the effect of the charge on the surface. To find the charge on the surface it is necessary to know i) the ion and electron fluxes, Γ_i and Γ_e, and how they depend on the electric field, and also ii) whether charge can leak from the surface of the trench into the substrate underneath the insulating layer (which is usually silicon) and if charge leaks away, how fast it leaks.

The issue of charging of the trench walls is primarily important when the walls are insulating. Suppose a trench is being etched into silicon and the sidewalls are covered in a perfectly insulating polymer film. The bottom must be kept clear of the film by ions in order for the etching to proceed. The silicon is usually a good enough conductor to be treated as a perfect conductor at a uniform electrostatic potential throughout the region near the trench. The electron and ion fluxes to the walls must be found as a first step. (The effect of the expected charging pattern is sketched in Fig. 7.4.)

Surface charging has been studied in Refs. [158], [169], and [170], and charging with more emphasis on charging-induced damage in Refs. [171]–[179]. There are currently differences of opinion between some of these authors as to the role of charging in causing damage. A review, as well as recent results, are given in [180] To analyze this issue in detail, the physical properties of the substrate must be known. For Si, some of the pertinent data are given in Refs. [29], [181], and [182].

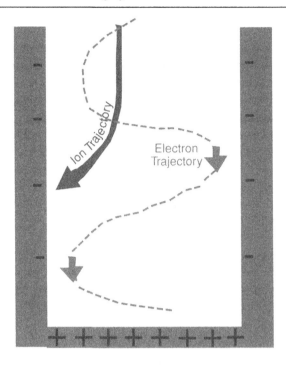

Fig. 7.4. Possible effects of trench charging on charged particle motion. If the sidewall charge is negative and the bottom charge is positive, ions will be attracted to the sidewalls (which will make the etching less directional) and electrons will be more likely to hit the trench bottom.

7.4.1 Electron and Ion Fluxes

This section discusses the fluxes in the sheath and the trench in more detail.

The Sheath Potential

The electrons that reach the surface of the wafer have overcome a large electrostatic potential V_{sh} that attempts to repel them, since any surface in contact with the plasma is showered by electrons until its negative charge and its potential build up enough to repel most of the electrons. The electrons reaching the surface are from the high-energy tail of the distribution. When these electrons that can reach the surface are in the center of the plasma they have kinetic energy of at least eV_{sh}, the energy they need to cross the sheath.

The magnitude of V_{sh} will be considered next. In most plasmas at low neutral pressure, the wall must be at a potential that confines those of the electrons with energies up to and a little beyond the ionization energy. Confinement of these electrons is necessary, so that enough of the electrons capable of causing ionization will stay in the plasma and create new ions and electrons to replace those reaching the walls. If the ionization energy is ε^* and the energy needed to escape to the wall is eV_{sh} then we expect $eV_{sh} \gtrsim \varepsilon^*$.

Loss Cone for Electrons Escaping to the Wall

The electrons reaching the wafer have energies that are (probably) high enough to ionize neutrals when they are at the center of the plasma. Their energies are also high enough for these electrons to escape to the wall. Both ionization and loss to the wall strongly distort the electron distribution. These are exactly the electrons for which the Maxwellian distribution is least likely to apply (with the possible exception of the electrons with the lowest total energy, which are trapped by the potential in the center of the plasma.) The easiest way to find the electron flux to the top surface of the wafer is to find the ion flux, provided that the surface is insulating so that the two fluxes are equal in steady state. This does not, however, tell us the electric field needed to equalize the ion and electron fluxes. First we shall consider finding the electron flux directly, to illustrate how the electrons behave in the sheath and in the trench.

Since the electrons in the trench are capable of reaching the wall, they are not necessarily isotropic, because the electrons moving toward the wall may not come back. The distribution of electrons is likely to be emptied out of particles moving in the direction away from the wall, if 1) their energy is high enough for them to have reached the wall and if 2) they are not likely to reflect off the wall.

Suppose the electrons were isotropic, with density n_λ, at a distance of about the electron elastic mean free path λ_e^{el} from the wafer. This distance is large enough for elastic collisions to largely remove the anisotropy caused by the surface. Even if the electrons have enough energy to reach the surface, they will not do so if their velocity is entirely parallel to the surface. The kinetic energy associated with $v_\perp$, the velocity normal to the surface, must be enough to overcome the potential barrier. Suppose that $\lambda_e^{\text{el}} > x_s$ (where x_s is the sheath thickness) and that the electrons are starting out isotropically distributed in velocity a distance λ_e^{el} away from the wall, which is outside the sheath. A kinetic energy of $\frac{1}{2} m_e v_\perp^2 > e V_{\text{sh}}$ is required if the electrons are to be able to reach the wall, or $v_\perp > v_\ell = \sqrt{2 e V_{\text{sh}}/m_e}$. The fraction of the electrons at any kinetic energy that can reach the surface corresponds to the fraction of the surface of a sphere of radius v_e such that $v_\perp$, which is the projection of v_e in the direction normal to the surface, is greater than v_ℓ. The subscript ℓ denotes "loss cone." The velocity v_ℓ corresponds to the value of the velocity toward the surface, when the velocity vector is at the edge of the loss cone at an angle θ_ℓ from the (inward) normal, and where $\cos \theta_\ell = v_\ell/v_e$ provided $v_\ell/v_e < 1$. $\theta_\ell = 0$ if $v_\ell > v_e$. Integrating over angle, we get the area of the part of the sphere having $\theta < \theta_\ell$:

$$A(\theta < \theta_\ell) = \int_0^\theta 2\pi v_e^2 \sin\theta \, d\theta = -2\pi v_e^2 \cos\theta \Big|_0^{\theta_\ell} = 2\pi v_e^2 [1 - \cos\theta_\ell].$$

(7.35)

The fraction of the sphere occupied by the angles $\theta < \theta_\ell$ is

$$f_\ell = A/4\pi v_e^2 = \frac{1}{2}[1 - \cos\theta_\ell] = \frac{1}{2}[1 - v_\ell/v].$$

(7.36)

This result means that at most half the electrons can get to the wall, the ones moving toward the wall, and this happens when $v_\ell = 0$. (v_ℓ cannot be negative unless the sheath attracts electrons, which is a possibility we shall not consider.) The region with $\theta < \theta_\ell$ is a "loss cone" for electrons, which as mentioned before is the reason for the subscript ℓ on θ_ℓ.

The fraction of electrons that can reach the wall is about f_ℓ, if we assume the electrons have their last collision at a distance of about λ_e^{el} from the surface. The flux per unit area of electrons to the wall, Γ_e^ω, is that part of the total flux per unit area of electrons that approached the sheath, Γ_e^{sh}, such that the electrons are inside the loss cone. $\Gamma_e^\omega = f_\ell \Gamma_e^{\text{sh}}$. Γ_e^ω is the flux that will enter unit area of the trench's top surface. Once the electrons enter the trench they will bounce around in the electrostatic potential and their direction will be partially randomized. Various descriptions of the behavior of the electrons in the trench might be applicable. The simplest is probably that the flux of electrons entering the trench distributes itself over the entire area of the trench's interior surface, which is energetically accessible to this group of electrons.

A more accurate prescription for determining the fate of the electrons would take account of their angular distribution as they enter the trench and allow them to move in the electric field inside the trench, to determine where they strike the wall if at all. This detailed calculation is probably not justified at present given all the uncertainties in the estimates we make.

Making Electron and Ion Fluxes Equal

If we know the ion flux to the wafer surface outside the trench, and if the surface is an insulator, we can set the total electron flux equal to the ion flux in steady state. The electron flux Γ_e is about $\frac{1}{4}n v_e$. The ion flux Γ_i is equal to the ion flux entering the sheath. The ion density is typically (very roughly) a tenth of the maximum ion density, at $x = x_s$, the point where the sheath meets the main plasma. To understand a particular experiment, the use of a measured value of the ion density at the sheath–plasma boundary would be appropriate. The ion velocity as it enters the sheath is mainly due to the ion falling though the electrostatic potential inside the plasma. The electron density usually obeys a Boltzmann density distribution, $n_e = n_0 \exp(e\Phi/k_B T_e)$, which implies $\Phi = \frac{k_B T_e}{e} \ln(\frac{n_e}{n_0})$, which in turn means the potential in the plasma is of the order of $\sim k_B T_e/e$. The ion energy as it enters the sheath is then $q\Phi \sim k_B T_e$, which is also (roughly) the mean electron energy, and the ion speed is $v_B = \sqrt{\frac{k_B T_e}{m_i}}$, with m_i the ion mass. (Here v_B is the "Bohm velocity.") The ion flux per unit area entering the sheath is $n(x = x_s)v_B$ and if the sheath area is roughly constant all the way across the sheath, this is the flux to the wafer. Setting $\Gamma_e = \Gamma_i$, and letting the electron speed v_e be the "thermal" electron speed v_e^{th}, we get

$$\frac{1}{4}n_e v_e^{\text{th}} = n_i(x = x_s)v_B \sim f_s n_0 v_B. \tag{7.37}$$

Here n_0 is the peak plasma density. f_s is the ratio of the typical ion density in the sheath to the peak density, and so $n_i(x = x_s) = f_s n_0$ with $f_s \sim 0.1$. This implies that the electron density at the wafer surface is $n_e \simeq 4 f_s n_0 v_B / v_e^{\text{th}}$.

The electron speed at the surface is about v_e^{th}, which is apparently paradoxical since the electrons lost energy eV_{sh} in reaching the wafer. It is a property of a Maxwellian distribution of particles that when the particles move into a region of higher potential energy, the density is lower but the particles still have the same distribution at the same temperature. The lowest energy particles are reflected, and the energetic particles lose kinetic energy, and so all the particles move down in kinetic energy in such a way as to keep the distribution the same. Then $v_{\text{B}}/v_e^{\text{th}} \simeq \sqrt{\frac{m_e}{m_i}}$ and n_e (at the wafer surface) $\simeq 4 f_s n_0 \sqrt{\frac{m_e}{m_i}}$.

To summarize our understanding of the electron and ion fluxes, we can say that the ions fall nearly straight toward the wafer surface. Various physical processes that deflect the ions from going straight down (thermal motion of neutrals and ions in the main plasma and collisions of ions) were discussed and their effects estimated. The electrons near the surface are repelled by negative charge on the surface, and only those electrons with both enough energy to reach the surface and with velocities within a loss cone pointing toward the surface can reach the surface. If the wafer surface is an insulator, then the fluxes must be equal at every point on the surface, although if the surface is conducting the fluxes only need to be equal when integrated over all the surface that is electrically connected. We are interested in a trench that is at least in part an insulator. Consequently, $\Gamma_e = \Gamma_i$ across much of the surface. The trench is so small that it is not likely to perturb the fluxes outside the trench and so $\Gamma_e = \Gamma_i$ at the trench mouth will hold and will guarantee that the net charge entering the trench per unit time is zero.

Particle Collection by Trench Sidewalls

As a simple model of the trench and the particle behavior in it, let the trench be rectangular with insulating sides and a conducting bottom. The potential of the bottom will be assumed to be no more positive than the surface of the wafer, and so ions are not repelled by it after they enter the trench. If ions entered the trench traveling straight down, in the absence of an electric field they would all reach the bottom. If the electrons all reached the bottom, then Γ_e and Γ_i would be equal at the bottom as well as at the top of the trench. The electrons are moving in all directions (except outward) when they enter the trench, however, so they will hit the sidewalls until the sidewalls charge up enough to repel additional electrons. Once the majority of the electrons could no longer reach the sidewalls then the electrons might nearly all be able to reach the trench bottom, if the potential due to the charge on the walls did not repel them from the trench altogether. The ions, however, would be strongly attracted to the walls by the potential of the walls, which must (in volts) be comparable to T_e (in eV) to repel the majority of the electrons.

The potential on the sidewalls will attract ions rather easily since a small deflection will bring ions in contact with the walls. Some ions are very close to the wall to begin with. Consider an ion that enters the top of the trench of depth t_{d} in the center of the opening of width t_{w}. For the ion to be able to hit the middle of the sidewall it needs an average velocity parallel to the wafer surface, $\bar{v}_{\parallel}$, that is large enough compared to the velocity normal to the surface, $v_{\perp}$, given by $\bar{v}_{\parallel}/v_{\perp} \sim t_{\text{w}}/t_{\text{d}}$. For it to avoid hitting the sidewall, it should have an average $\bar{v}_{\parallel}$ during its flight down the trench so that $\bar{v}_\parallel / v_\perp \lesssim t_w / 2 t_d$. Its energy associated with $v_\parallel$ would thus be about $\frac{1}{2} m v_\parallel^2 \lesssim \frac{1}{2} m v_\perp^2 (\frac{t_w}{2 t_d})^2$ or very roughly $T_e (t_w / t_d)^2$. If the trench has an "aspect ratio" $t_d / t_w = 5$ then the wall potential need only be about $T_e / 25$ for the wall to collect many of the ions.

These estimates indicate that the insulating walls need only be at a negative potential of a few percent of T_e to collect a large fraction of the ions. This potential will repel some electrons. Exactly what fraction of T_e causes Γ_e to equal Γ_i will depend on issues such as whether the charged particles bounce off the sidewalls. Ions hitting at a glancing angle and bouncing could have an important effect, even if they are neutralized when they bounce, since they will convert into energetic neutrals, which will hit the bottom and may help to etch the bottom. Analysis of the gas-phase processes has thus again indicated that the biggest determinant of reactor performance may be the surface processes. The surface roughness and composition will have a considerable effect on what happens to charged particles hitting the surface at different energies and angles. The surface's role is not well understood at present. Suitable experiments to probe particle–surface interactions in reactor-relevant conditions are clearly needed to clarify these issues.

7.4.2 Trench Electrostatics

It is useful to examine the implications of an approximate solution of Laplace's equation,

$$\nabla^2 \Phi = \frac{\partial^2 \Phi}{\partial x^2} + \frac{\partial^2 \Phi}{\partial y^2} + \frac{\partial^2 \Phi}{\partial z^2} = 0. \tag{7.38}$$

The solution we will use corresponds to uniform charge densities on the walls and the bottom of a rectangular trench and a total charge inside the trench equal to zero. The potential is

$$\Phi = \frac{\beta}{2} (y^2 - x^2). \tag{7.39}$$

Inside the trench the coordinates are (x, y) with x being horizontal and y vertical and with the origin at the center of the top of the trench. The left wall is at $x = -t_w / 2$, the right wall is at $x = t_w / 2$, and the bottom is at $y = -t_d$. The components of the electric field $\mathbf{E}$ are

$$E_x = -\frac{\partial \Phi}{\partial x} = \beta x \tag{7.40}$$

and

$$E_y = -\frac{\partial \Phi}{\partial y} = -\beta y. \tag{7.41}$$

At the sidewalls

$$E_x (x = \pm t_w / 2) = \pm \beta t_w / 2. \tag{7.42}$$

This field points into the surface, at both sidewalls. The field normal to the surface ends on negative surface charges (or volume charges inside the walls), either on top of the insulating layer or behind it. If the terminating charge is all surface charge on the top of the insulator, then the charge per unit area of the sidewall σ_s is given by

$$\sigma_s = -\varepsilon_0 E_{normal} = -\varepsilon_0 \beta t_w/2. \tag{7.43}$$

At the bottom of the trench, $E_y = \beta t_d$, which points upward, and $\sigma_b = \varepsilon_0 \beta t_d$. Since the charge on the bottom is positive, electric field lines can start on the bottom and end on the negative charges on the trench walls. The area of the bottom, using a unit length in the z direction, is t_w and so the charge on the bottom is $Q_b = \sigma_b t_w = \varepsilon_0 \beta t_d t_w$. The area of a sidewall is t_d and so the charge on one sidewall is $Q_s = \sigma_s t_d = -\varepsilon_0 \beta t_w t_d/2$. Combining the charge on both sidewalls and the bottom gives a total charge of zero.

This solution is not totally realistic but has some useful features. If the charges on the walls and the bottom were in reality spread out uniformly, this solution for Φ would be quite accurate. This situation would be most likely to occur in a trench with all its sides being insulators. If the bottom is open to the silicon we expect the potential at the bottom to be constant. This solution has $\Phi(y = -t_d) = \beta(t_d^2 - x^2)/2$, which is nearly constant since x is assumed small compared to t_d, but it does give rise to a nonzero E_x at $y = -t_d$. For a conducting trench bottom the positive charges on the trench bottom will be attracted toward the sidewalls until $\Phi(y = -t_d)$ is uniform. (The positive charges on the silicon would be holes, or else donor impurities exposed by electrons that retreated from the region, depending on the doping of the silicon and on the strength of the electric field.)

7.4.3 Ion Motion in the Trench

Ion motion in the trench can be calculated using this potential. The ion enters the top of the trench at some location x_0 and with its velocity **v** approximately straight down. Its acceleration in the x direction is given by

$$m\frac{d^2x}{dt^2} = eE_x = e\beta x. \tag{7.44}$$

This equation has solution $x = Ae^{\gamma t} + Be^{-\gamma t}$, where $\gamma \equiv (e\beta/m)^{1/2}$. The initial conditions at $t = 0$ are $x(t = 0) = x_0 = A + B$ and $\frac{dx}{dt}|_{t=0} = 0 = \gamma(A - B)$. These imply $A = B = \frac{x_0}{2}$ and so

$$x = x_0(e^{\gamma t} + e^{-\gamma t})/2. \tag{7.45}$$

We further assume that the y velocity is constant (thus the ions move fast enough so that their y velocity is not affected much). Then $y = -vt$ and

$$x = x_0(e^{-\gamma y/v} + e^{\gamma y/v})/2. \tag{7.46}$$

As the ions move down the trench it seems reasonable to expect the exponentially growing term to dominate, but to confirm this we need to know the value of γ/v.

Now $\gamma^2/v^2 = \frac{e\beta}{mv^2}$, and $E_x = \beta x$. We expect E_x is a fraction of T_e/t_w. (The potential set up by the plasma is usually of the order of T_e, and if the length over which it changes is t_w then $E \sim V/t_w \sim T_e/t_w$.) Then $\beta \sim \alpha T_e/t_w^2$, where α is smaller than one, and since $\frac{1}{2}mv^2 \sim eT_e$, $\gamma^2/v^2 \sim \alpha/t_w^2$. Although α is small, when $(-y)$ becomes considerably larger than t_w the exponentially growing term can dominate. To simplify matters we set $x = \frac{1}{2}x_0 e^{-\gamma y/v}$ when solving for the depth y_f at which the ion hits the wall. When the ion hits the wall, $x = t_w/2 = \frac{1}{2}x_0 e^{-\gamma y/v}$, which gives

$$y_f = -\left(\frac{v}{\gamma}\right)\ln\left(\frac{t_w}{x_0}\right) = \left(\frac{v}{\gamma}\right)\ln\left(\frac{x_0}{t_w}\right). \tag{7.47}$$

Now $x_0 \lesssim t_w/2$ and so $y_f < 0$. This expression is not appropriate for small y_f because the second exponential term was neglected and also because the number of ions striking the wall at small y_f will depend on the ions' angular distribution, which was ignored.

This expression shows how the ions distribute themselves over the sidewalls. A fraction of the ions equal to $\Delta x/t_w$ enter the trench in each strip of width Δx of the mouth of the trench from $(x_0 - \Delta x/2)$ to $(x_0 + \Delta x/2)$. These ions go into a width Δy on the wall, where

$$\Delta y = \left(\frac{v}{\gamma}\right)\left[\ln\left(\frac{x_0 + \Delta x/2}{t_w}\right) - \ln\left(\frac{x_0 - \Delta x/2}{t_w}\right)\right]$$

$$= \left(\frac{v}{\gamma}\right)\ln\left(\frac{x_0 + \Delta x/2}{x_0 - \Delta x/2}\right) \simeq \left(\frac{v}{\gamma}\right)\ln\left(1 - \Delta x/x_0\right) \simeq -\frac{v\Delta x}{\gamma x_0}, \tag{7.48}$$

or $\frac{\Delta y}{\Delta x} \simeq -\frac{v}{\gamma x_0}$ provided $\Delta x \ll x_0$. If the ions enter the trench with a large x_0 (i.e., near the edge of the trench) they go to a "small" y (i.e., near the top of the trench) in a "small" range of y values, whereas if x_0 is small they are spread out into a "large" range of y values at a "large" y (deep in the trench). This is to be expected; this expression shows quantitatively how it happens.

The flux of ions entering the mouth of the trench is $n_{se}v_B$ particles per square meter per second (where n_{se} is the plasma density at the sheath edge). The strip of width Δx and of unit length in the z direction is hit by $n_{se}v_B\Delta x$ particles per second, and these particles hit the wall spread over an area of dimensions Δy in depth by unit length in the z direction. Dividing the number per second $n_{se}v_B\Delta x$ by the area $|\Delta y|$ gives the number hitting per unit of the sidewall area per second,

$$n_{se}v_B\Delta x/|\Delta y| = n_{se}v_B\gamma x_0/v. \tag{7.49}$$

Alternatively, since $x_0 \simeq t_w \exp(\gamma y/v)$, the number per unit area of sidewall per second is

$$n_{se}\gamma\left(\frac{v_B}{v}\right)t_w\exp\left(\frac{\gamma y}{v}\right), \tag{7.50}$$

where, as usual, y is negative. The ion flux to the trench wall is seen to decay exponentially with depth.

7.4.4 Electron Motion in the Trench

For electrons entering the mouth of the trench the potential we have been using is confining in the x direction (that is, it keeps electrons off the walls) and tends to pull the electrons down into the trench in the negative y direction. The electron velocity at the trench mouth is expected to be roughly uniformly distributed in angle, for all negative v_y that take the electron into the trench.

The electron enters the trench at $x = x_0$ with horizontal component of velocity v_{x0} and the component of velocity down the trench $v_y = v_{y0}$. The equation of motion in the x direction is

$$m_e \ddot{x} = -eE_x = -e\beta x. \tag{7.51}$$

This is a simple harmonic oscillator equation, with frequency $\omega_{e\tau} = (e\beta/m_e)^{1/2}$. Its solution is

$$x = A\cos\omega_{e\tau}t + B\sin\omega_{e\tau}t. \tag{7.52}$$

The initial conditions at $t = 0$ when the electron enters the trench are $x = x_0 = A$ and $\dot{x} = v_{xo} = \omega_{e\tau}B$. Therefore

$$x = x_0\cos\omega_{e\tau}t + (v_{x0}/\omega_{e\tau})\sin\omega_{e\tau}t. \tag{7.53}$$

The maximum x the electron can reach is $x_m = (x_0^2 + (v_{x0}/\omega_{e\tau})^2)^{1/2}$, and if $x_m > t_w/2$ the electron will hit the sidewall. If $x_m < t_w/2$ the electron will bounce from side to side until it reaches the bottom of the trench (at least in this potential, since the potential will not reflect the electron back out of the trench mouth).

The electron equation of motion in the y direction is

$$m_e \ddot{y} = -eE_y = e\beta y, \tag{7.54}$$

which has exponentially varying solutions, given by

$$y = A'\exp(-\omega_{e\tau}t) + B'\exp(\omega_{e\tau}t). \tag{7.55}$$

The initial conditions at $t = 0$ are $y(t = 0) = 0 = A' + B'$ and $\dot{y} = -v_{y0} = \omega_{e\tau}(-A' + B')$. These imply $B' = -A'$ and $v_{y0} = 2\omega_{e\tau}A'$ so that

$$y = \frac{v_{y0}}{2\omega_{e\tau}}\{\exp(-\omega_{e\tau}t) - \exp(\omega_{e\tau}t)\}. \tag{7.56}$$

This is zero at $t = 0$ but becomes negative for $t > 0$. The equation for x can in principle be solved for the time when the electron hits the wall by setting $x = \pm t_\omega/2$, solving for the time for both signs of x, and choosing the smallest positive time when the electron hits the wall. This is a more complicated procedure than is justified by the accuracy of the overall model. Instead, an estimate of the time equal to one half of the period of the oscillation $T = \frac{2\pi}{\omega_{e\tau}}$ can be used; the electron hits the wall in the first half-period, that is, in $T/2 = \pi/\omega_{e\tau}$, or not at all. The range of y values where the electron hits the wall is from zero to y_m, where

$$y_m = \frac{v_{y0}}{2\omega_{e\tau}}\left\{\exp\left(-\omega_{e\tau}\frac{\pi}{\omega_{e\tau}}\right) - \exp\left(\omega_{e\tau}\frac{\pi}{\omega_{e\tau}}\right)\right\} = \frac{v_{y0}}{2\omega_{e\tau}}(e^{-\pi} - e^{\pi}), \quad (7.57)$$

or

$$y_m \simeq -14\,v_{y0}/\omega_{e\tau}. \quad (7.58)$$

This range of depths is rather large and implies that electrons with enough energy in the vertical direction can travel long distances down the trench before they hit the wall. These electrons may hit the trench bottom before they hit the wall, even for trenches with a high aspect ratio, $A = t_d/t_w$, the trench depth divided by the trench width. The tendency of the electrons to be accelerated rapidly downward helps to keep them away from the wall and to decrease the need for a large repulsive potential to keep them off the wall.

To estimate the wall potential, the maximum distance electrons go before hitting the wall, y_m, can be set equal to the trench depth t_d and this can be used to find the electric field:

$$|y_m| \simeq 14\,v_{y0}/\omega_{e\tau} = t_d. \quad (7.59)$$

The initial kinetic energy T_y associated with the y motion is $T_y = \frac{1}{2}m_e v_{y0}^2 \simeq \frac{m_e}{2}$ $(\frac{\omega_{e\tau}t_d}{14})^2$, or $(\omega_{e\tau}t_d)^2 \simeq 400 T_y/m_e$. From the definition of $\omega_{e\tau}$ this is $(\omega_{e\tau}t_d)^2 = \frac{e\beta t_d^2}{m_e}$ and so $\beta \simeq 400\,T_y/(et_d^2)$. The quantity (T_y/e) is equal to the initial kinetic energy of the y motion in electron volts.

The potential of the wall V_w at $t_w/2$ relative to the center of the trench at $x = 0$ is

$$V_w = \frac{\beta}{2}\left(\frac{t_w}{2}\right)^2 = 50\left(\frac{t_w}{t_d}\right)^2 T_y(\text{eV}). \quad (7.60)$$

The aspect ratio is $A = t_d/t_w$; therefore $V_w \simeq 50\,T_y(\text{eV})/A^2$. The potential of the trench bottom V_b relative to $y = 0$ is much larger. $V_b = -\beta t_d^2/2 \simeq -200\,T_y(\text{eV})$. If the trench aspect ratio were $A = 10$ and $T_y(\text{eV}) \sim 1/2$ eV then $V_w = 0.25$ volts, but $V_b = -100$ volts. These values seem extreme. Smaller values of A would give less exaggerated voltages.

A second relationship, between $\omega_{e\tau}$ and the amplitude of the oscillation x_m, can be obtained by assuming that at $x_0 = 0$ the x velocity v_{x0} is comparable to v_{y0} so that $x_m \sim v_{y0}/\omega_{e\tau}$. If $v_{y0} \simeq \omega_{e\tau}t_d/14$ then $x_m \sim t_d/14$, which implies that for a relatively short trench, an electric field that is strong enough to pull typical electrons all the way down the trench before they hit the wall may also be strong enough to hold them off the wall altogether. Short in this case means with a small enough depth t_d such that $x_m \sim t_d/14 \lesssim t_w/2$, and the electron does not hit the wall. Then the aspect ratio is $A = \frac{t_d}{t_w} \lesssim 7$. For longer trenches the charge density on the wall is spread out making E_x weaker.

Exercises

1.   For a trench with an initial shape that is rectangular, how will the shape evolve if the etching is purely chemical and the density of chemicals is uniform inside the trench? Is this realistic? Why might the trench evolve differently in reality?

2.   Repeat numerically the calculation of the rate of change of the angle made by the side of the trench, $\frac{d\theta_s}{dt}$. When the sections are not equal in length, instead of using x_2 to find the etch rate of the vertical section, use $x_{13} = \frac{1}{2}(x_1 + x_3)$. Compare the numerical results to the analytic results.

3.   For the analytic electric field given in this chapter, $E_x = \beta x$ and $E_y = -\beta y$, numerically solve the equations of motion for ions entering a trench at the top of the trench at $y = 0$. Start the ions with $v_y = v$ and $v_x = 0$. Launch equal numbers of ions per unit length of the top of the trench. Plot the number of ions hitting unit length of the sidewall, as a function of depth. Compare the results to the analytic expressions for the number per unit length obtained in the text.

4.   For the same electric field as in the previous question, solve the equations of motion for electrons that enter the trench with $v_y = -v_{yo}$ and $v_x = v_{xo}$. Examine several orbits, and comment on the analytic calculations and the assumptions that went into them. In particular, are the estimates of y_m and V_w reasonable?

5.   What are the most important unknown factors that make it difficult to set up an accurate model of trench evolution during etching? Explain why they are important. How could you set up a model of trench evolution to test the role of these factors, (i.e., to check how important they are) and to see if experimental results can be reproduced, for "reasonable" guesses at numerical values for parameters?

8

Physical Description of the Plasma

In this chapter we go into more detail in describing how calculations of plasma properties may be done. Analytic models, the role of experiments in model building, and computational models are considered. We provide the basis for a mathematical description of the transport and of the density of the plasma in a plasma reactor. Next, we discuss experimental design and its role in model building. We then turn to a survey of the most widely applicable computational methods for describing plasmas.

8.1 Analytic Plasma Models

We begin this section with a review of solutions to simple diffusion equations in one and two dimensions. We turn, after that, to a one-dimensional description of the plasma. The plasma is divided into the main plasma and the sheath. The main plasma is quasineutral, whereas the sheath, which is the region of strong electric fields near the chamber wall, is positively charged since it has relatively very few electrons in it. The main plasma is also divided into a "presheath" and an interior region. The presheath is the part of the main plasma, next to the sheath (so at the outside of the main plasma) which is responsible for most of the acceleration of the ions before they enter the sheath. The division into main plasma and presheath is arbitrary, but it might be expected that the presheath would be of the order of an ion mean free path in thickness, since that is the distance within which the ions can expect to accelerate freely so that is the distance in which their acceleration must be achieved.

We then consider two-dimensional transport in a plasma. The emphasis in this section is on the way to use physical reasoning to obtain reasonable predictions.

In the plasma interior, an ambipolar diffusion equation holds. A few analytic cases of the solution of a "simple" diffusion equation are worth examining, since they provide some insight into more realistic problems.

8.1.1 1D Diffusion, Step Source

The "step source" is a source that is zero in part of the range and constant in the rest: $S = S_0$ for $0 \leq x < a$, and $S = 0$ for $a \leq x < b$. The steady-state diffusion equation for $S = 0$ is

$$-D\frac{d^2n}{dx^2} = 0. \tag{8.1}$$

Integrating once we get

$$\frac{dn}{dx} = c. \tag{8.2}$$

Integrating again we have

$$n = e + cx. \tag{8.3}$$

In the region where $S = S_0$, the steady-state diffusion equation is

$$-D\frac{d^2n}{dx^2} = S_0, \tag{8.4}$$

or

$$\frac{d^2n}{dx^2} = -\frac{S_0}{D}.$$

Integrating once, we get

$$\frac{dn}{dx} = f - \frac{S_0 x}{D} \tag{8.5}$$

and integrating again yields

$$n = g + fx - \frac{S_0 x^2}{2D}. \tag{8.6}$$

The boundary conditions we shall use are that $n = 0$ at $x = 0$ and $x = b$ and that n and $\frac{dn}{dx}$ must be the same at $x = a$, where the two solutions must match.

The condition $n(x = 0) = 0$ means that $g = 0$, whereas $n(x = b) = 0$ means that $e + cb = 0$ and so $e = -cb$.

Then for $0 \leq x < a$,

$$n = x\left[f - \frac{S_0 x}{2D}\right], \tag{8.7}$$

and for $a < x \leq b$,

$$n = c[x - b]. \tag{8.8}$$

At $x = a$ these must be equal. Setting $n_0 \equiv \frac{S_0 a^2}{2D}$, we get

$$af - n_0 = c[a - b]. \tag{8.9}$$

The derivatives $\frac{dn}{dx}$ must also be equal;

$$f - \frac{2n_0}{a} = c. \tag{8.10}$$

Substituting for c, we have

$$af - n_0 = \left(f - \frac{2n_0}{a}\right)(a - b), \tag{8.11}$$

which gives

$$n_0 = -b \left(f - \frac{2n_0}{a} \right). \tag{8.12}$$

Therefore

$$f = n_0 \left(\frac{2}{a} - \frac{1}{b} \right). \tag{8.13}$$

The density from $x = 0$ to $x = a$ is

$$n = xn_0 \left(\frac{2}{a} - \frac{1}{b} - \frac{x}{a^2} \right) \tag{8.14}$$

and from $x = a$ to $x = b$ is

$$n = n_0 \left(1 - \frac{x}{b} \right). \tag{8.15}$$

8.1.2 2D Diffusion with "Simple" Sources

In this section we continue to review the analytic solution of the diffusion equation. The main reason for doing this here is that this calculation is perhaps the most sophisticated of the textbook analyses worth pursuing. For greater realism we would need to go to an asymptotic analysis or, more likely, to a numerical treatment. Detailed solutions of nonlinear transport equations, based on a Fourier method (which is at the heart of the material in this section), have been performed, however; see Refs. [183] and [184].

In two spatial dimensions (x, y), with a rectangular boundary at $x = 0$ and $x = a$, $y = 0$ and $y = b$, and $n = 0$ on the boundary, the steady-state diffusion equation with constant diffusion coefficient D becomes

$$-D\nabla^2 n = -D \left(\frac{\partial^2 n}{\partial x^2} + \frac{\partial^2 n}{\partial y^2} \right) = S. \tag{8.16}$$

This equation can be solved for a given $S(x, y)$ by "separation of variables." Two forms for S will be considered: S equal to a constant, S_0, and S varying exponentially with x, $S = S_0 e^{-x/\delta}$. The exponential variation is supposed to reflect a situation that might occur in a high-pressure ICP, where the rf fields drop exponentially with distance.

The solution is found in two parts: i) a "complete" solution n_h of the homogeneous equation, which has $S = 0$, and ii) a "particular" solution n_p of the full equation, with the real S, where the solution for n_p does not need to match the boundary conditions. If the two solutions are added, the overall solution will satisfy the full equation, since

$$-D\nabla^2 n_h = 0 \tag{8.17}$$

and

$$-D\nabla^2 n_p = S \tag{8.18}$$

and so

$$-D\nabla^2(n_{\rm h} + n_{\rm p}) = S. \tag{8.19}$$

The homogeneous solution $n_{\rm h}$ will have enough unknown coefficients in it to allow the overall solution, $n = n_{\rm h} + n_{\rm p}$, to be made to match the boundary conditions.

The homogeneous solution we start with is the same in both cases and is well known. We choose the separation constant k in the separation of variables method so that we get

$$n_{\rm h} = \sum_k \sin kx [A_k e^{ky} + B_k e^{-ky}]. \tag{8.20}$$

Terms involving $\cos kx$ have been left out to make $n_{\rm h}$ be zero at $x = 0$. We choose $k = \frac{m\pi}{a}$ so that $n_{\rm h}$ is zero at $x = a$. We can set $n_{\rm h}$ to zero at these edges because we shall be able to find particular solutions $n_{\rm p}$ that are zero at the edges in x.

The particular solution we shall use when $S = S_0$, of

$$-D\nabla^2 n = S_0, \tag{8.21}$$

is a one-dimensional solution (which goes to zero at $x = 0$ and $x = a$):

$$n_{\rm p} = -\frac{S_0}{2D}x(x-a). \tag{8.22}$$

However, when $S = S_0 e^{-x/\delta}$, we find by twice integrating

$$-D\frac{d^2 n}{dx^2} = S_0 e^{-x/\delta} \tag{8.23}$$

that

$$n_{\rm p} = A + Bx - \frac{\delta^2 S_0}{D}e^{-x/\delta}. \tag{8.24}$$

The choice of A and B that makes $n = 0$ at the edges gives

$$n_{\rm p} = n_0\left[1 - e^{-x/\delta} - \frac{x}{a}\left(1 - e^{-a/\delta}\right)\right]; \tag{8.25}$$

$$n_0 \equiv \frac{\delta^2 S_0}{D}. \tag{8.26}$$

Note that when δ is very large, these two results for $n_{\rm p}$ become the same.

The values of $n_{\rm p}$ at the edges in y are the same (since $n_{\rm p}$ does not depend on y) and must be combined with $n_{\rm h}$ to give zero at $y = 0$ and $y = b$. This means that $n_{\rm h}$ must be the same at $y = 0$ and $y = b$.

This implies that

$$A_k + B_k = A_k e^{kb} + B_k e^{-kb}, \tag{8.27}$$

and thus

$$B_k = A_k \frac{(1 - e^{kb})}{(e^{-kb} - 1)} = A_k e^{kb}, \tag{8.28}$$

and if we define $a_k = A_k + B_k$, which is the coefficient of $\sin kx$ at $y = 0$ or $y = b$, then

$$A_k = \frac{a_k}{e^{kb} + 1}. \tag{8.29}$$

We chose $k = \frac{m\pi}{a}$ and so for $m = 1$ only half of a period fits into the range $x = 0$ to $x = a$. We will pretend that the boundary condition is antisymmetric about $x = a$; that is, the solution for n_{p} (and hence also for n_{h}) from $x = a$ to $x = 2a$ is a negative mirror image of that from $x = 0$ to $x = a$, with the mirror being at $x = a$. This is helpful, in that it makes the average of the boundary value of n_{p} (and similarly also n_{h}) be zero.

To find the coefficients we write $n(y = 0)$ as

$$n(x, y = 0) = n_{\mathrm{p}} + \sum_k a_k \sin kx = 0. \tag{8.30}$$

We multiply by $\sin k'x$, where $k' = \frac{m'\pi}{a}$, and integrate from $x = 0$ to $x = 2a$. Since n_{p} is antisymmetric about $x = a$, even values of m' will give zero when $\int_0^{2a} n_{\mathrm{p}} \sin k'x\, dx$ is performed. Only odd values of m' need be kept. For m' odd, $\int_0^{2a} n_{\mathrm{p}} \sin k'x\, dx = 2\int_0^a n_{\mathrm{p}} \sin k'x\, dx$ (where the upper limit was changed to get the integral on the right). The other term in the equation gives $\int_0^{2a} a_k \sin kx \sin k'x\, dx$ $= aa_k$, provided $k' = k$.

The two ionization profiles we considered were $S = S_0$ and $S = S_0 e^{-x/\delta}$. The solution for a_k when the source is constant ($S = S_0$) is

$$a_k = -\frac{4}{ak^3} \frac{S_0}{D}. \tag{8.31}$$

When $S = S_0 e^{-x/\delta}$,

$$a_k = -\frac{2n_0}{a} \frac{e^{-a/\delta} + 1}{k(1 + k^2\delta^2)}, \tag{8.32}$$

with $n_0 = \frac{\delta^2 S_0}{D}$ and, in both cases, $k = \frac{m\pi}{a}$, but m, the integer multiplying $\frac{\pi}{a}$ must be odd. (See Fig. 8.1.) Again, the two results agree in the limit when δ is very large.

8.1.3 The Presheath

The physical picture of a plasma confined in a vacuum vessel involves a narrow charged region called a "sheath," near the wall and, next to the sheath, a wider quasineutral "presheath" in which ions accelerate to the Bohm velocity before entering the sheath. Suppose, for the moment, that the presheath is somewhat narrow. Two useful consequences follow from the presheath being narrow:

a) The flux in the presheath is roughly constant since most of the ionization occurs in the main plasma.
b) The flux in the presheath moves straight outward as if in one spatial dimension.

The density and the electric field in the presheath can be estimated, assuming

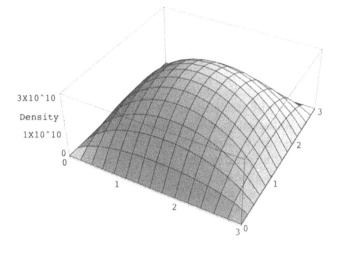

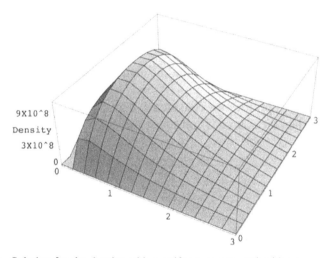

Fig. 8.1. Solution for the density with a uniform source, and with an exponentially varying source.

i) the electrons obey a Boltzmann distribution

$$n_e = n_0 \exp(e\Phi/k_B T_e) \qquad (8.33)$$

(and in the quasineutral regions outside the sheath the electron density is very nearly equal to the ion density) and

ii) the ion velocity is the velocity the ions pick up between collisions. This means the mean ion velocity is given roughly by $\frac{1}{2} m v_i^2 = eE\lambda_i/2$, which is the energy picked up in half a mean free path.

Assumption i) means that

$$E = -\frac{d\Phi}{dx} = -\frac{k_B T_e}{e n_e}\frac{dn_e}{dx}$$

and $n_e = n_i$. Therefore the ion speed is given by

$$v_i^2 = -\frac{e\lambda_i}{m_i}\left(\frac{k_B T_e}{e n_i}\frac{dn_i}{dx}\right) = -\frac{k_B T_e \lambda_i}{m_i}\frac{1}{n_i}\frac{dn_i}{dx}. \tag{8.34}$$

The ion flux per unit area is $\Gamma_i = n_i v_i$ and so $v_i^2 = (\Gamma_i/n_i)^2$. Setting the two expressions for v_i^2 equal to each other, we get

$$\left(\frac{\Gamma_i}{n_i}\right)^2 = -\frac{k_B T_e \lambda_i}{m_i}\frac{1}{n_i}\frac{dn_i}{dx}, \tag{8.35}$$

or

$$n_i\frac{dn_i}{dx} = \frac{1}{2}\frac{dn_i^2}{dx} = -\frac{m_i \Gamma_i^2}{k_B T_e \lambda_i}. \tag{8.36}$$

This has the solution

$$n_i^2 = n_p^2 - \alpha x, \tag{8.37}$$

with $\alpha \equiv 2m_i \Gamma_i^2/(k_B T_e \lambda_i)$. Here n_p is the density at $x = 0$, which, in this calculation, will be the edge of the presheath adjacent to the plasma interior.

If we require that $v_i^2 = v_B^2 \simeq 2k_B T_e/m_i$ at the sheath edge, then at the sheath edge, where $x = x_s$, $n_i = n_s = \Gamma_i/v_B$. (v_B is the Bohm velocity. x_s denotes the position of the sheath edge, but when studying the sheath it was measured from a different origin than now, since $x = 0$ is currently the edge of the presheath which is next to the main plasma.) Since α can be written as $\alpha = 4\Gamma_i^2/\lambda_i v_B^2$, then $n_p^2 - n_s^2 = \alpha x_s$ becomes

$$n_p^2 - \left(\frac{\Gamma_i^2}{v_B^2}\right) = \frac{4\Gamma_i^2}{v_B^2}\frac{x_s}{\lambda_i} \tag{8.38}$$

and the presheath thickness is

$$x_s = \frac{\lambda_i}{4}\left(\frac{v_B^2 n_p^2}{\Gamma_i^2} - 1\right). \tag{8.39}$$

Now $\Gamma_i = n_i v_i$, so at the point where ions enter the presheath $\Gamma_i = n_p v_p$, where v_p is the speed of the ions when they enter the presheath, and

$$x_s = \frac{\lambda_i}{4}\left(\left(\frac{v_B}{v_p}\right)^2 - 1\right). \tag{8.40}$$

8.1.4 The Plasma Interior

The region of the plasma further in than the presheath will now be treated. In what follows the coordinate system is shifted so n_0 is now the density at the center of the plasma, at the new origin of coordinates $r = 0$, $z = 0$.

The behavior of the plasma further in than the presheath is difficult to treat in general. The ionization rate depends in a relatively complex way on the method of heating, the neutral pressure, and so on. If we assume the inelastic electron mean free path λ_e^{inel} is large, the electrons' elastic mean free path λ_e^{el} is nevertheless usually considerably smaller than the system size. The effective length for inelastic collisions λ_e^* is the net distance particles travel while diffusing a total distance of about λ_e^{inel} in N_s steps of length about λ_e^{el}. Now $\lambda_e^* = \sqrt{N_s}\lambda_e^{\text{el}}$ whereas from the definition of N_s, $\lambda_e^{\text{inel}} = N_s\lambda_e^{\text{el}}$ and so $\lambda_e^* = \sqrt{\lambda_e^{\text{el}}\lambda_e^{\text{inel}}}$. If λ_e^* is also larger than the system size, the most energetic electrons should be able to diffuse throughout the interior of the plasma (where the electrostatic potential Φ is low, so they have enough kinetic energy to ionize) before they have an inelastic collision. In other words, if λ_e^* is large the ionization rate S might be relatively uniform (at least in the region where the electric field is comparatively weak). If λ_e^* is small we expect ionization to be localized near where the electrons gain kinetic energy, for instance from an rf field, or by "falling" to a region where they have lower potential energy.

If S is constant ($S = S_0$) and we consider a problem in one dimension, then in steady state the flux of ions per unit area, Γ_i, is equal to all the ions created per unit area per second "upstream," $\Gamma_i = \int S dx = S_0 x$. Using the same arguments as before, we obtain

$$\left(\frac{\Gamma_i}{n_i}\right)^2 = -\frac{k_B T_e \lambda_i}{m_i}\frac{1}{n_i}\frac{dn_i}{dx}. \tag{8.41}$$

Therefore

$$\frac{dn_i^2}{dx} = -\frac{2m_i S_0^2 x^2}{k_B T_e \lambda_i} \tag{8.42}$$

and

$$n_i^2 = n_0^2 - \beta x^3, \tag{8.43}$$

where $\beta \equiv 2m_i S_0^2/(3k_B T_e \lambda_i)$. Then the electric field is equal to

$$E = -\frac{k_B T_e}{2en_i^2}\frac{dn_i^2}{dx} = \frac{k_B T_e}{2e}\frac{3\beta x^2}{n_0^2 - \beta x^3} = \frac{m_i S_0^2 x^2}{\lambda_i e(n_0^2 - \beta x^3)} \tag{8.44}$$

$$\simeq m_i S_0^2 x^2/\lambda_i e n_0^2, \tag{8.45}$$

assuming n_i^2 is relatively uniform in the interior of the plasma.

This calculation should be contrasted with the "simple" diffusion equation that was solved in Section 8.1. We assume the diffusion is "ambipolar," and the ion flux is calculated as a response to an electric field. The electric field is found in the same way, in both calculations. The difference is in how the ion flux is found from the electric field. This "long mean free path" calculation uses energy conservation to show that v_i^2 is proportional to $E\lambda$. The "diffusive", short mean free path, picture of plasma transport, uses an ion drift velocity $v_d = \mu_i E$. This difference in the dependence of the ion flow velocity on E causes the difference in the answers obtained.

To match the solution inside the plasma onto the solution in the presheath, we should have continuity in n_i and $\frac{dn_i}{dx}$. In the plasma interior,

$$n_i^2 = n_0^2 - \beta x^3, \qquad x < x_\mathrm{p}, \tag{8.46}$$

and in the presheath,

$$n_i^2 = n_\mathrm{p}^2 - \alpha(x - x_p), \qquad x \geq x_\mathrm{p}, \tag{8.47}$$

where x_p is the lower edge of the presheath.

Setting $n_\mathrm{p}^2 = n_0^2 - \beta x_\mathrm{p}^3$ ensures continuity of n_i at $x = x_\mathrm{p}$. The derivatives are made equal by setting

$$\frac{dn_i^2}{dx} = -3\beta x_\mathrm{p}^2 = -\alpha. \tag{8.48}$$

However, since $\beta = 2m_i S_0^2 / 3k_\mathrm{B} T_e \lambda_i$ and $\alpha = 2m_i \Gamma^2 / k_\mathrm{B} T_e \lambda_i$ then by setting the fluxes equal, $\Gamma = S_0 x_\mathrm{p}$, we have $3\beta x_\mathrm{p}^2 = \alpha$ automatically, without an additional constraint on x_p. This is because continuity of dn_i^2/dx follows from the equality of the fluxes that was imposed.

The division into presheath and main plasma was arbitrary; the only requirement was that the source was unimportant in the presheath so that Γ was constant (although this is not necessary in a presheath in general). For the change in Γ to be small we may need the presheath to be narrow. If we recall that the width of the presheath was denoted as x_s, then n_i^2 at the sheath edge of the presheath is

$$n_i^2 = n_\mathrm{p}^2 - \alpha x_s = \left(n_0^2 - \beta x_\mathrm{p}^3\right) - \alpha x_s$$
$$= n_0^2 - \alpha(x_\mathrm{p}/3 + x_\mathrm{s}). \tag{8.49}$$

We have called the "presheath" the region with no source in it, so this result is valid for large x_s provided $S = 0$ in the region denoted as the "presheath."

8.1.5 Two-Dimensional Transport

To extend this type of analysis to two dimensions is considerably more difficult. If the plasma is much longer in one direction than the other, then the short direction could be treated as one-dimensional as a first approximation. We will then try to improve the approximation later.

To illustrate the use of a series of approximations like this, we treat the case of a short, fat ICP with a relatively short electron effective mean free path for ionization λ_e^*. Then the ionization takes place near the coil where the hot electrons are produced. The ionization rate will be assumed to be of the form

$$S = S(z) \exp(-(r - r_c)^2 / 4\Delta r^2). \tag{8.50}$$

Then the lowest-order solution for the density, $n_0(z)$, will be found using the one-dimensional analysis given above, with z as the sole variable, to find $n_0(z)$. This first stage of analysis will be independent of radius, and we will ignore the exponential factor in S but treat S as $S(z)$. This calculation will not be performed here; it will be assumed that it can be done, for the appropriate form of $S(z)$ – perhaps using one of the analytic treatments given above.

Once $n_0(z)$ is obtained, the lowest-order density will be obtained by putting back the exponential factor, so that

$$n(r, z) = n_0(z) \exp(-(r - r_c)^2/4\Delta r^2). \tag{8.51}$$

The radial transport is responsible for filling in the density at $r = 0$. We shall now look for the way that the radial transport could cause this.

If the above expression accurately represents the density, then the radial electric field can be found from it:

$$E_r = -\frac{k_B T_e}{e n_i} \frac{dn_i}{dr} = \frac{(r - r_c)}{2(\Delta r)^2} \frac{k_B T_e}{e}. \tag{8.52}$$

In this radial direction, the spatial scale is often much larger than the ion mean free path. Thus it might be appropriate to assume that the ion transport is diffusive and that ambipolar diffusion, with ambipolar diffusion coefficient D_a, will occur. The radial flux is

$$\Gamma_r = -(D_a \nabla n)_r = D_a n_0(z) \frac{(r - r_c)}{2(\Delta r)^2} e^{-(r-r_c)^2/4(\Delta r)^2}. \tag{8.53}$$

(It might have been more straightforward to find this flux from $\Gamma_r = n_i \mu_i E_r$, using E_r given above. The ambipolar flux of ions is actually almost entirely due to ion drift in the electric field.) The divergence of this radial part of the flux is $\nabla \cdot \Gamma = \frac{1}{r} \frac{\partial}{\partial r}(r \Gamma_r)$. The divergence of the flux is the number leaving per unit volume per second. In this calculation, $-\nabla \cdot \Gamma$ will be used to give the number entering unit volume per second:

$$\nabla \cdot \Gamma = \frac{D_a n_0(z)}{2r(\Delta r)^2} \frac{\partial}{\partial r} [r(r - r_c) \exp\{-(r - r_c)^2/4(\Delta r)^2\}]$$

$$= \frac{D_a n_0(z)}{2r(\Delta r)^2} \left[2r - r_c - \frac{r(r - r_c)^2}{2(\Delta r)^2}\right] e^{-(r-r_c)^2/4(\Delta r)^2}. \tag{8.54}$$

If the ionization is localized close to the coil then Δr is small and $r/\Delta r$ is large. Consequently, $\nabla \cdot \Gamma$ can be approximated as

$$\nabla \cdot \Gamma = -\frac{D_a n_0(z)}{4(\Delta r)^4} (r - r_c)^2 e^{-(r-r_c)^2/4(\Delta r)^2}. \tag{8.55}$$

The negative of this divergence of the radial flux will be treated as an extra source term in the one-dimensional transport in the z direction. It will be included in the calculation of the next approximation to $n_0(z)$. The extra source is just $S_{\text{div}} = n_0(z)R(r)$, with

$$R(r) = \frac{D_a(r - r_c)^2}{4(\Delta r)^4} \exp(-(r - r_c)^2/4(\Delta r)^2).$$

When this is added to the original source, the procedure for finding the density at a given radius as a function of z can be performed as before. The importance of the extra source term obtained above is that it is positive and so it shows how the plasma tends to fill in the otherwise hollow density profile near $r = 0$.

One notable aspect of the result for the one-dimensional case is that the electric field is very weak in the interior of the plasma. If the electric field is only large at large distances from the center of the plasma, what this really means is that the presence of a boundary is necessary for a substantial field to exist. The field points toward that boundary, and the plasma near the boundary can again be treated as nearly one dimensional. The one-dimensional result obtained in 8.43, (with a constant ionization rate $S = S_0$) for the electrostatic potential Φ is

$$\Phi(z) = -\frac{k_B T_e}{2e} \ln\left(\frac{n^2}{n_0^2}\right) = -\frac{k_B T_e}{2e} \ln\left(1 - \frac{\beta z^3}{n_0^2}\right). \tag{8.56}$$

The equivalent result in the radial direction is

$$\Phi(r) = -\frac{k_B T_e}{2e} \ln\left(1 - \frac{\beta r^3}{4n_0^2}\right), \tag{8.57}$$

where $r = 0$, $z = 0$ is in the center of the plasma. In the radial direction the analysis is changed slightly by the geometrical effect of the radius. The radial flux per unit area Γ_r equals the total number of particles produced per second per unit length in z within a radius r, divided by $2\pi r$. The total number produced per unit length in z per second is $\int_0^r 2\pi r S dr = \pi r^2 S_0$ and this equals $2\pi r \Gamma_r$ so that $\Gamma_r = r S_0/2$. An extra factor of $\frac{1}{2}$ has been introduced into Γ_r as compared to the result in Cartesian coordinates, but since β varies as Γ^2 an extra factor of $\frac{1}{4}$ appears.

Combining terms and setting $\Phi(r, z) = \Phi(r) + \Phi(z)$ in an attempt to get a 2D version of this is not justified, because the flux of particles is what drives the variation in density and potential. If we combine terms, we 'double count' the flux.

$$\Phi(r, z) = -\frac{k_B T_e}{2e} \ln\left(1 - \frac{\beta z^3}{n_0^2}\right)\left(1 - \frac{\beta r^3}{4n_0^2}\right)$$

$$\simeq -\frac{k_B T_e}{2e} \ln\left(1 - \frac{\beta}{n_0^2}\left(z^3 + \frac{r^3}{4}\right)\right). \tag{8.58}$$

Similarly, the estimated density is

$$n^2 \simeq n_0^2 - \beta\left(z^3 + \frac{r^3}{4}\right). \tag{8.59}$$

However, this result is based on an assumption that, to lowest order, the transport processes are one-dimensional. This may be partially justifiable near a wall where r^3 is large and z^3 small, or vice visa. If we compute $\nabla \cdot \Gamma$ from this, the result will be too large, as a result of the double-counting.

8.2 Experimental Design

To understand the performance of a plasma reactor we have several tools available. For clarity we can divide the tools we use into

 i)   experiments,
 ii) complex computational models of the reactor, and
 iii) simple analytic models of the reactor.

The effective use of any of these tools requires insight obtained from the others. To be specific, some of the many interconnections include the following facts: a) Clearly any useful models must agree with whatever experiments are available. Experiments on plasma reactors do not usually provide a complete characterization of the reactor, however, so more information is needed. b) Complex computer models are in a sense numerical experiments. They can provide more detail, since they are in principle completely characterized – it is possible, although not always easy, to extract all the data as to what is happening. Numerical experiments can thus test analytic models. c) Because numerical experiments are frequently difficult to understand, simple analytic models are still helpful to interpret them. The complex models often suggest analytic models. d) The other reason why analytic models are necessary, when using computational models, is that straightforward use of standard numerical methods frequently fails to capture the real plasma behavior. It is often essential to build into a calculation the properties that we know the physical system should exhibit.

So far the case has been made in these examples for comparing computational models and analytic models with experiment and with each other, to improve each of them. The understanding of the reactor, which is obtained from all classes of models, should also be fed back into the design of experiments. In this way all three of the elements available are used in an iterative procedure. Each provides insight, which allows improvement on how the others are implemented. (See Fig. 8.2.)

The method of experimental design that will be presented here is based on exploiting models of a system in designing experiments on the system and using experiments to test the models, all in the most effective way possible.

The essence of much of the work that has been done on experimental design by statisticians is in a sense straightforward. The starting point for this discussion is that it is accepted that experiments on complex systems should be designed systematically and using all the available information to optimize the design. It then follows that first the independent variables in the experiment should be examined to see which are expected to be important; the design of the experiment should reflect the list of variables which are identified and should allow a determination of which variables are actually significant. At the same time, the data should be examined for indications that important variables have been omitted. This stage is essentially a prescreening or pretest.

Particularly during a pretest it is likely that many variables will be involved in the experiment. When many independent variables are present, it is important to extract information about as many of them as possible, from each experiment that is performed. The need to obtain as much information as possible, from a limited number

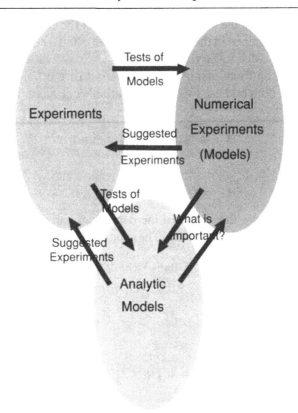

Fig. 8.2. The relationship among experiments, numerical experiments, and analytic models. Ideally, all of these should be used iteratively and in combination, to explain the behavior of a physical system.

of experiments, leads to the use of factorial design of experiments. In a factorial design the independent variables are typically assigned "low" and "high" values. All the possible permutations of low and high settings of all the variables are examined in a full factorial design. In a partial factorial an attempt is made to extract the same information using fewer experiments. A partial factorial design requires an appropriate choice of the permutations of low and high settings. This choice of permutations must sample the low and high settings of each individual variable, without testing all possible values of the other variables, for each setting of a given variable.

For example, with three independent variables (which can be used to define axes in a Cartesian coordinate system), low and high values of each variable define the ends of the ranges of each variable. Taken all together the low and high values define the faces of a rectangular box. The possible permutations of low and high correspond to the corners of the box. A full factorial design would usually involve experiments corresponding to each of the eight corners. A partial factorial might involve experiments corresponding to four corners, chosen so that none of the corners employed were adjacent to each other.

The importance of independent variable one, x_1, can be tested by appropriately averaging the values of a dependent variable, y. Two averages are obtained. These averages are intended to represent in the first case the value at the center of the face having $x_1 = x_{1\text{Low}}$ and in the second case the value at the center of the face having $x_1 = x_{1\text{High}}$. The difference in the averages gives an estimate of the variation in the dependent variable y, due to varying x_1 from its low value to its high value.

The first average is found using two corners out of the four employed in the partial factorial. At both of these two corners, variable one has a "low" value. These might be at

$$C_1 = (x_{1\text{Low}}, x_{2\text{Low}}, x_{3\text{High}})$$

and at

$$C_2 = (x_{1\text{Low}}, x_{2\text{High}}, x_{3\text{Low}}).$$

The second average uses the values of the dependent variable at the two corners where variable one has a "high" value. These corners might be at

$$C_3 = (x_{1\text{High}}, x_{2\text{Low}}, x_{3\text{Low}})$$

and at

$$C_4 = (x_{1\text{High}}, x_{2\text{High}}, x_{3\text{High}}).$$

Then we estimate

$$y(x_{1\text{Low}}) = 1/2(y(C_1) + y(C_2))$$

while

$$y(x_{1\text{High}}) = 1/2(y(C_3) + y(C_4)).$$

The reduction in the number of experiments by going from a full factorial to a partial was from eight to four. With three independent variables this is helpful but not essential. With much larger numbers of independent variables, partial factorials are much more beneficial.

In the pretest stage, analytic models would have suggested which variables should affect the outcome of the experiments. In later stages, analytic models will imply the sensitivity of the dependent variables to the independent variables, allowing the "low" and "high" values to be chosen appropriately. The change in the dependent variable(s) in going between the low and high values of each independent variable should probably be of the same order, if the effects of one independent variable are not to be swamped by the effects of another. This may not be possible in practice, however.

The results of the experiments will test the analytic models. In the absence of accurate, or sufficiently detailed, analytic models the experiments will frequently be used to provide the basis for finding response surfaces. The term response surface is used here to mean a fitted relationship between a dependent variable and the independent variables. The response surface can provide a point of contact with computer experiments and a guide for improving analytic models. It can be a useful step in the process of developing a picture of the reactor, although it should not be the end of the process, since it does not by itself imply a physical basis for understanding the reactor. A response curve would be useful in a control strategy for the reactor, to show how adjusting an independent variable can push the dependent variable in a desirable direction. It will, however, be considerably more useful to understand why the response curve has the shape it does, since such an understanding can be used to extrapolate to other situations whereas the response curve only allows interpolation between known states of the experiment.

Similar remarks apply to the use of neural nets, which can "learn" how a system will respond. Neural nets can often interpolate effectively from previous experience from a system, although neural nets can make dramatic errors since it is not uncommon for them to be used inappropriately. Unfortunately, they do not directly provide any understanding of the behavior of the system they represent. As a result it is difficult to anticipate when the correlations they have found, between the dependent variables and the independent variables, will fail to hold.

8.2.1 Measurements of Plasma Properties and Model Development

Probably the greatest difficulty in understanding plasmas is caused by the lack of comprehensive data on plasma behavior. Plasma diagnostics tend to disturb the plasma (as in the case of probes) or to give partial information (as in most current spectroscopic measurements, which might yield the line-average density, along a particular chord, of a single species in the plasma). Mass spectroscopic measurements can be used to analyze the mix of ions leaving the plasma at some point, but these measurements do not directly yield information on conditions in the interior of the plasma. Cross sections and other fundamental data are also not well known, and information on surface processes is even harder to obtain.

It is sometimes claimed that plasma theory lags behind plasma experiments. Reality is more complex than this. The capabilities of existing plasma models can be illustrated if we imagine an ideal plasma reactor experiment, where all the possible data about the plasma are known with arbitrarily high accuracy. Fundamental data on cross sections and reaction rates are also supposed to be known in this idealization.

The construction of a global model of all the aspects of the reactor plasma would probably involve sub models, one for each of the major components identified previously. The global model would function by iteratively employing all of the sub models, probably in succession. With existing modeling capabilities it would be possible to feed into each sub model the measured values for all the necessary input data for that sub model and obtain results that were very close to reality. Suppose all the models were used in conjunction with each other but without frequent contact with the measured values. Then, with existing models, errors would probably tend to accumulate and lead to results that were only moderately accurate. However, if this ideal experiment was available for comparison with models, meaningful tests of the models would be possible. The testing process would permit production of accurate first-principle calculations, which did not need input from experiments during the iterations. The primary obstacle to development of an accurate and reliable modeling capability is the lack of accurate data on fundamentals such as cross sections and on enough of the aspects of some relevant discharge(s), which would allow tests of whether or not the various sub models work properly.

In what follows we will review the computational methods which are available for use in the different sub models.

8.3 Computational Models of Plasmas

The development is underway of large-scale computational models of plasmas. These employ numerical solutions of the equations that arise from all the diverse sub models we have introduced. Some of the techniques we shall present are straightforward to use. Nevertheless, most of these computational models are difficult to implement and require very large amounts of computer resources. In fact computer speed and memory create major limitations on the calculations that can be performed. Because these methods consume considerable amounts of resources it is vital to be aware of their advantages and disadvantages and especially to have a realistic set of goals for the work. Those unfamiliar with the capabilities of the models often have expectations that are incommensurate with the models. However, the goals of those doing the modeling frequently only extend as far as demonstrating some measure of "agreement" between a particular model and an experiment. There is a presumed implication that, if the model "agrees" with some experiment, then a predictive capability has been demonstrated for the model. If there is no theoretical basis for believing the model to be valid, it is not justifiable to use the model to extrapolate trends outside the region where it is known to reproduce experimental results. Models are, however, used outside their region of validity and their use is justified as being somehow expedient. The problems with doing with this are compounded by the fact that the supposed "agreement" may be no more than with regard to trends, as opposed to precise agreement as to numerical values. In the absence of a solid foundation for belief in the applicability of a model, the model is no better than a fitted relationship between variables, and the model may in fact be a very bad fit to the relationship.

The conclusions from this are that 1) one should only use models that one has good reason to believe are correct and that 2) before starting to use a model, one should have a clear idea as to what it will show that is useful. A scientific goal for the model might be to show that a particular physical process, when included in the model, is capable of reconciling the model and experiment. Once a credible demonstration of agreement has been achieved, then the model can be used for an engineering purpose, such as to design an improved reactor.

In what follows we introduce several simulation techniques. The main emphasis is on kinetic simulations, which are introduced after a brief discusion of fluid models. The simulation methods are illustrated in the context of particular calculations. The scientific goals of the calculations can be discussed, once we have discovered what each technique is capable of doing. Most of these methods, as applied to rf discharges, were compared in Ref. [185]. The main conclusion seems to have been that kinetic methods agree with each other better than they agree with fluid models.

8.3.1 Fluid Models

Fluid equations were discussed at some length in Chapter 4 and in this chapter. The book by von Engel [45] is still extensively cited on this subject. The solution of fluid equations by numerical methods is a very substantial subject; a thorough treatment, including a discussion of the Scharfetter–Gummel method, as well as software tools to assist in setting up simulations, is given in Ref. [29] and references therein. (The automation of setting up the fluid equations is considered in Refs. [29] and [186].)

One of the major issues addressed in Ref. [29] is that of mesh generation – see also Refs. [187]–[189]. The use of irregular meshes to solve equations of the form

$$\nabla \cdot \Gamma = S \qquad (8.60)$$

– that is, equations involving the divergence of a vector, which include most of the equations we use in a fluid description – was described in Ref. [29]. In particular, "box integration" was introduced. In box integration, the definition of the divergence as the flux leaving a volume divided by that volume (in the limit when the volume is very small) is used to contruct the divergence in a natural way on an "arbitrary" mesh. Software tools were provided to make it easy to set up both the mesh and the equations which are to be solved.

There are a variety of works on numerical methods, mostly in the context of solving finite difference equations and/or fluid equations, that are helpful or interesting for various reasons [190–205]. Some recent examples of fluid plasma modeling include Refs. [206] and [207]. The use of the Scharfetter–Gummel method with finite elements occurred much later than its use with finite differences – see Ref. [208].

8.3.2 Monte Carlo Simulations of Plasmas

This and the next several sections consider "kinetic" simulations of plasmas. In this context, kinetic means that information is generated about how many particles are moving at each velocity. Perhaps the simplest kinetic simulation to understand is the so-called Monte Carlo simulation, and so we start with this. It will also provide illustrations of concepts needed in other methods. (An interesting comparison of Monte Carlo and other kinetic methods is given in Ref. [209].)

The name "Monte Carlo" is used in connection with computer simulations of physical systems to indicate that the simulation makes use of random numbers. Choosing random numbers is also an element in games of chance, which is the link to Monte Carlo. Scientific computer languages allow the programmer to call a random number generator within a program. The random numbers used in what follows are uniformly distributed between zero and one, unless otherwise indicated.

Suppose that an event has a probability p of occurring. A call to the random number generator returns a number r between zero and one. The value of r may be used to indicate whether or not the event "occurred" on any given trial. If the random number $r < p$ (p being the probability) then the event will be said to have occurred; if $r > p$ the event will be said to have not occurred. For example, if the equations of motion of a particle are being integrated with a time step Δt and the collision frequency is ν, then the probability of a collision in each time step is approximately $\nu \Delta t$, provided that $\nu \Delta t \ll 1$. If $p = \nu \Delta t = 0.1$ then collisions should occur during 10% of the time steps. Since the random number r is uniformly distributed between zero and one, in general r is less than p in a fraction of the trials equal to p. In our example with $p = 0.1$ then $r < p$ that is, r is less than 0.1, in 10% of trials. If, for instance, a call to the random number generator results in $r = 0.09$, then the particle "had a collision," and so its velocity will be adjusted to allow for the effects of the collision, whereas if the result of the call was $r = 0.11$ it "did not collide" and so its velocity is unchanged by collisions in that time step.

Monte Carlo simulations of plasmas are often straightforward to set up. Interpreting the results of the simulations can be subtly difficult because the Monte Carlo "particles" that are followed are only a sample from the "real" particles. The Monte Carlo particles' behavior is, in turn, a sample of the real behaviors. Some important groups of particles are very few in number compared to the total number of particles. Thus it will not be easy for the sample of particles that are followed to adequately sample those small groups. The importance of some rare types of particles is a particular concern in plasma simulation, as hinted at by all of our discussion of the role of the tail of the electron distribution. An important early paper that used the method in semiconductor simulation gives a clear account [210].

It is helpful to discuss the method in concrete applications, which serves the secondary purpose of illustrating the physics occurring in that particular plasma. The next section is devoted to describing the Monte Carlo simulation of a particular dc discharge. More complex discharges will be handled, using this and other methods, later.

Monte Carlo Simulation of DC Discharges

The earliest use of Monte Carlo simulation in application to discharge plasmas that we are aware of was in the work of Tran, Marode, and Johnson [211]. A dc discharge was modeled having a Cathode Fall (CF) adjacent to the cathode, followed by a Negative Glow (NG). The electric field used in the simulation is based on measurements, and so it is not updated during the calculations. This electric field is not necessarily consistent with the charge density, which is found from the behavior of the Monte Carlo particles. The electric field in the CF is frequently assumed to vary linearly, from its maximum magnitude at the cathode to nearly zero at the other side of the CF. The NG electric field can probably be neglected in the calculation for many purposes. Electrons are pushed away from the cathode by the CF field. As they move away from the cathode they accelerate and pick up kinetic energy until they have enough kinetic energy to ionize neutrals. Since the electrons produced by the ionization are also pushed away from the electrode, they also gain energy and ionize neutrals. In this way they contribute to the "avalanche" of electrons. Ions produced by this ionization fall to the cathode. The ions striking the cathode cause a (usually) smaller number of "secondary" electrons to be emitted from the cathode. These secondary electrons are the seeds of the avalanche, and thus are responsible for sustaining the discharge.

To describe the electrons using a Monte Carlo method, a computer simulation is used in which the equations of motion of a sample of electrons are integrated over time. The time integration may move all the electrons simultaneously, with time step Δt. Alternatively (and since the calculation is not being done self-consistently, so it is not necessary to update the electric field using the charged particle density), the simulation particles can be followed one at a time until a maximum time has elapsed or until the electron being followed escapes from the system. If electrons are followed individually, the first electron considered is probably a secondary electron, which leaves the cathode after an ion hits the cathode. This electron accelerates in the strong electric field in the CF and undergoes elastic and inelastic collisions. It will probably cause several ionization events. The electrons produced by ionization are also followed, including all the "generations" of particles. When the avalanche of electrons due to the one electron have all escaped or the electrons have all run out of time, another electron leaving the cathode is followed, and so on. The number of ions produced in ionization events is also tracked. These ions all hit the cathode. Suppose one electron is emitted from the cathode for every N^{av} ions that hit the cathode. Then each secondary electron's avalanche needs to create N^{av} ions, if the system is in steady state.

8.3.3 Particle In Cell Simulations

Particle In Cell (PIC) methods usually involve following large numbers of simulation particles simultaneously [55]. The particle density, mean velocity, or other average quantities are estimated from the sample, which is provided by the simulation particles. The PIC method allows the forces that the particles exert on each other to be found from these averages and to be included self-consistently in the calculation. At each time step the particles' behavior (position; velocity) is updated, followed by the force, then the next step is taken, and so on. PIC methods have been applied extensively to simulation of processing plasmas [58, 59, 212–217].

In PIC simulations of plasmas, the density of charged particles is the most important average quantity since it determines the electrostatic potential Φ through Poisson's equation,

$$\nabla^2\Phi = -\rho/\varepsilon. \tag{8.61}$$

Here ρ is the "free" charge density and ε is the dielectric constant; $\rho = e(n_i - n_e)$ with e the magnitude of the electronic charge and n_i (n_e) the ion (electron) number density. The quantities n_i and n_e are estimated from the number of simulation particles in each "cell" of the simulation mesh.

The first problem with PIC plasma models is that n_i and n_e fluctuate artificially. This is because the number of simulation particles is too small to make the expected variation in number be very small compared to the number. If there are of order $N_s \sim 10^6$ simulation particles, representing a total number of real particles of perhaps 10^{12} particles, then each simulation particle represents 10^6 real particles. If there are 100 cells on the mesh, in a one-dimensional simulation, and simulation particles are placed in the cells randomly then the expected number in a cell is $N_c = 10^4$, but the fluctuation in N_c is about $\sqrt{N_c} = 100$, or 1% of N_c. The number of real particles, represented by this fluctuation in the number of simulation particles, would be a fluctuation of 10^8 real particles in the cell. A fluctuation of 10^8 real particles would be very unlikely to happen, since the mean number in the cell is about 10^{10}. The fluctuation in this would be $\sqrt{10^{10}}$, or 10^5. A fluctuation of 10^8 real particles would cause very large electric fields if it were not compensated by particles of the opposite charge. Those electric fields would attract particles of the opposite charge into the cell and would largely neutralize the excess charge.

Nevertheless, these fluctuations in the simulation cause unphysical oscillations in density and electric field, which in turn cause unphysical heating of the particles and especially of the electrons, which respond fastest to the electric field. Attempts are made to remove the effects of fluctuations by using time-averaged densities in Poisson's equation and other similar schemes. It is not clear how successful the averaging is.

Limits on the Time Step in a Self-Consistent Calculation

Because in a self-consistent simulation the electrostatic potential Φ is updated at each time step, using Poisson's equation, phenomena can take place that could not occur in a Monte Carlo simulation where Φ is not changed over time. Various types of oscillations and waves are now permitted. A particularly important instance is the plasma oscillation, which is an oscillation in which electrons are displaced relative to the ions and are pulled back by the electric field that is set up. If collisions are absent then the oscillations are not (strongly) damped. If collisions are frequent the displacement decays exponentially. In either case the time for the oscillation or the decay sets a characteristic physical time scale, which to some extent limits the size of the time step that can be used in the simulation. Unless a somewhat sophisticated technique is used to update the particle positions and the electric field, the time step must be much less than the characteristic physical time. Even with a more complex method, the time step can only be a few times the characteristic physical time. The need for the small step is to ensure that the physical process is resolved. To resolve an oscillation it is necessary (according to the Nyquist criterion) to take several (at least two) time steps within each single period of the oscillation.

To see how the plasma oscillation occurs, and to calculate its frequency, a slab of plasma can be imagined, with a uniform ion density inside the slab and zero ion density outside. If the electrons are not displaced their density is equal to the ion density at each point. If the electrons are displaced by a distance x to the right, then a net positive charge is exposed on the left of the slab in a layer of thickness x, whereas on the right a negative charge also of thickness x extends out of the slab. The charge density in the positive layer is $\rho = ne$, where n is the number density. The charge per unit area of the layer, viewed from the side, is thus $\rho x = nex$. The electric displacement D points from the positively charged layer, through the slab, to the negative layer and is equal to the charge per unit area, $D = nex$. The electric field E is $E = D/\varepsilon = nex/\varepsilon$ and so the force on an electron is $-eE = -ne^2x/\varepsilon$. This force equals the mass of the electron times its acceleration, if there are no collisions:

$$-\frac{ne^2x}{\varepsilon} = m\frac{d^2x}{dt^2}. \tag{8.62}$$

In a collisional case, the effect of the electric field is to drive the electrons at a velocity $v_e = -\mu_e E$, where μ_e is magnitude of the electron mobility and the negative sign of the electron mobility is shown explicitly. Since $v_e = \frac{dx}{dt}$ and

$E = nex/\varepsilon$ then

$$\frac{dx}{dt} = -\frac{\mu_e nex}{\varepsilon} = -\frac{x}{\tau_d}. \qquad (8.63)$$

In the collisionless case, the equation is a simple harmonic oscillator equation of the form

$$\frac{d^2x}{dt^2} = -\frac{ne^2}{m\varepsilon}x = -\omega_p^2 x. \qquad (8.64)$$

This has the solution

$$x = A\cos\omega_p t + B\cos\omega_p t, \qquad (8.65)$$

where $\omega_p \equiv (ne^2/m\varepsilon)^{1/2}$ is the "plasma frequency."

It is interesting to solve the collisional case, which gives an exponential decay: if we have frequent collisions, the appropriate equation is

$$\frac{dx}{dt} = -\frac{x}{\tau_d} \qquad (8.66)$$

which has the solution

$$x = x_0 e^{-t/\tau_d}, \qquad (8.67)$$

where $\tau_d \equiv \varepsilon/\mu_e ne$ is the dielectric relaxation time. The time τ_d is the characteristic time when the velocity is found from a mobility, which is the case when fluid equations are used.

In a typical PIC simulation the time step is limited by the plasma period, $T_p \equiv 2\pi/\omega_p$; the time step Δt is typically about a tenth of T_p if a straightforward explicit scheme is used. This limit is in addition to the limit set by the collision frequency v, which is that $v\Delta t \ll 1$. An implicit scheme may allow a time step of a few times T_p.

Advantages and Disadvantages of PIC

The advantages of PIC models are that they are very flexible and they describe the motion of the simulation particles very accurately, in the sense that there is no uncertainty in the position or velocity of each simulation particle, which there is in the mesh-based models described later.

However, since PIC models of discharge plasmas require that many simulation particles are followed, with quite short time steps, they are likely to be computationally intensive. Implementation of a PIC model is not entirely straightforward because of the need for careful handling of the electric fields, either by some form of averaging or by use of an implicit scheme, or both. Even though N_s, the number of simulation particles, is large, it is not large enough to resolve the particle behavior in the presence of the density variations that occur in some discharges. If the density changes by a factor of 10^8 between spatial cells or velocity cells, then even if almost all N_s simulation particles are in the high density cell there can only be about $10^{-8} N_s$ simulation particles in the low-density cell. For present-day simulations with $N_s \sim 10^6$ this means we have a 1% chance of finding even one simulation particle in the low-density cell. There has not as yet been implemented in a PIC model of a discharge an entirely satisfactory model of Coulomb collisions. This may be rectified in future, using methods similar to the treatment of Coulomb collisions employed in other "kinetic" plasma simulations [82].

8.3.4 PIC Simulation of an ECR Reactor

The application of a PIC model to an Electron Cyclotron Resonance (ECR) discharge [218] will illustrate how the model works in a case where most of the limitations of the PIC technique do not pose serious problems. The microwave fields that heat the plasma have a frequency of 2.45 GHz. Because the plasma density does not respond directly to this oscillation by oscillating significantly in turn, the ECR plasma resembles a dc discharge. There is no oscillating sheath and, further, the dc sheaths have only modest voltages across them; consequently, the density gradient across the sheath is probably smaller than in most dc and rf discharges. The accurate description of the sheath is perhaps less vital than in many reactors since, unlike in many dc and rf discharges, it is not responsible for heating the plasma. Lack of spatial resolution is thus probably not a major problem in this case. However, resolution of the high-energy regions of velocity space, which have few particles, is still a cause for concern. Coulomb collisions are not expected to play a major role because the electrons are relatively hot (the high temperature decreases ν_{ee}, the electron–electron collision frequency) and because other processes (for example, the ECR heating) change the electrons' energy and have a similar but more rapid effect, compared to Coulomb collisions, on electrons of all energies.

A problem common to nearly all plasma simulations is that the time scale for ions to escape from the plasma is much longer than the time scale on which the electrons respond to changes or settle down to a steady state. The evolution of the plasma in time could be thought of as being a sequence of small changes in the ion density, after each of which the electrons rapidly adjust to the new ion density, as does the electric field. This difference in time scales leads to a difficulty, because the time steps must be short enough to resolve the fastest electron behavior. However, the simulation must continue long enough to allow the slow ions to reach steady state. Problems like this, in which very large variations in the time scales must be handled by the calculation, are said to be numerically "stiff."

The ability of the electrons to "relax" rapidly can be exploited in the simulation of a discharge [219]. The ions and electrons were both followed using a short time step, suited to the electrons, until the electrons and the electric field had reached a steady state. The ions were then followed by themselves, using a longer time step more suited to the ion behavior, and using the ionization rate and electric field found previously from the full simulation where the electrons were also followed. This ions-only simulation was continued until the ion density had changed by an amount of the order of 10%, by which time the electron behavior might have changed slightly. The "full run" was then restarted and continued until the electrons and electric field again settled down. Then the ions-only simulation run was done again, and so on until a true steady state was reached. This procedure is much faster than integrating only on the electron time scale, since in this approach, steps based on the electron time scale were only used for a small fraction of the total elapsed "real" time. The rest of the "real" time was integrated using a time step based on the ion time scale, which is much longer.

The electric field used in the ions-only simulation could be the same field that the full PIC generated at the end of the full PIC stage of the simulation. A more accurate way to find the electric field during the ions-only stage is to assume i) the electron density in the interior of the plasma is equal to the ion density found from the ions-only simulation and ii) the electron density obeys a Boltzmann relation, with the electron temperature T_e being found during the full PIC. These assumptions allow the electrostatic potential Φ to be found, using $\Phi = -(k_B T_e / e) \ln(n_e / n_0)$ and $n_e = n_i$.

8.3.5 Hybrid Simulations

The type of numerical simulation of a plasma (or other complex system) that is at the same time the most accurate and effective might be a hybrid simulation. Ideally a hybrid simulation exploits the advantages of the different simulation methods and avoids the disadvantages of each, by using different types of simulation at different stages of the overall calculation. A hybrid scheme could for instance be one that uses a kinetic description of the electrons and a fluid description of the ions. In this type of hybrid, different physical levels of description are employed in distinct physical situations. A second type of hybrid could use different physical levels of description in a single physical situation. The PIC model of the ECR discharge described previously provides such an example. In the first stage of the calculation, the ions and electrons were both followed using the PIC approach, which is fully kinetic. In the next stage the ions were still followed using particle simulation but the electrons were included in the calculation in a much simpler way. The electrons were assumed to obey the Boltzmann relation in this stage of the hybrid simulation.

To describe the relative merits of different hybrids it will be useful to have a classification scheme for the various possibilities that exist. Focusing on the plasma model, as opposed to the treatment of the neutral species or the wall chemistry, we can use fluid (F) or kinetic (K) descriptions or a combination of these, FK, for both species, ions (i) and electrons (e). The PIC model described above uses a fully kinetic ion model (K_i). It also uses a kinetic model of the electrons part of the time but a very simple fluid model of the electrons at other times, and so the electron model will be denoted FK_e. The overall model is a K_i–FK_e hybrid. The extension of this scheme to neutrals is obvious. Most simulations use a fluid description of the neutrals (F_n), but some use a kinetic approach to neutral transport (K_n).

The most developed hybrid plasma model is Kushner's, [220–223], which usually operates as a F_i–FK_e hybrid and has been applied to many different problems. The electron model differs in a significant way from that in the PIC model, even though both are denoted here as FK_e. The PIC model described above alternates between a fluid model of the electrons and a kinetic model of the electrons. Kushner's approach subdivides the electron model differently; it uses a fluid model to find the density of electrons and a kinetic (Monte Carlo) model to find the ionization rate S. This reflects the same reasoning, outlined in an Chapter 4, that says that a fluid description of the majority of electrons, which are less energetic, is adequate. The less energetic electrons determine the density of the electrons. A fluid description of the most energetic electrons, which cause the ionization, is not adequate. Instead, a kinetic description of the most energetic electrons is needed to find the ionization rate. To distinguish this permutation a new notation, K_S, will be used, which will denote Kinetic Source. This hybrid is then F_i–F_e–K_S. A model that estimates the ionization rate from the density of electrons and from their temperature would be said to be a Fluid Source (F_S) model.

In the low-pressure plasmas used in plasma processing it is rarely justifiable to use a fluid approach to find the ionization rate; consequently, the F_S approach will be approximate at best. As Kushner points out, the combination F_e–K_S has several advantages, which will be described next, but one problem with it should also be kept in mind.

The advantage of the F_i–F_e–K_S hybrid is that each module is reasonably accurate and efficient. Computational effort is only expanded where it is necessary. For example, the kinetic description of electrons is most necessary for the source rate and less so for the bulk electron density. Hence the F_e–K_S description of electrons is appropriate. The advantages of such a combination were discussed in an earlier chapter. A fluid ion model is often accurate enough and hence F_i is employed. Different levels of description of the neutrals, F_n and K_n, have been used in this hybrid.

The problem is that the fluid electron (F_e) aspect of the hybrid is not entirely compatible with the kinetic source (K_S). The power deposition implied by the fluid model will be somewhat different from the power deposition used by the kinetic calculation of the source. This is because the fluid fluxes are determined by the mobility and the diffusion coefficient (or at least using similar, "fluid" concepts). The mobility cannot reproduce the effect of the real electric field, because it only uses the local electric field. The electron flux at any point in a low-pressure plasma actually depends on the electric fields at all points in the plasma. The diffusion coefficient suffers from similar problems. The FK_e combination in the PIC approach above uses a fully kinetic electron model, at least some of the time, so provided the simulation converges there is no problem with incompatibility of aspects of the electron model.

More than one hybrid has been applied to modeling inductively coupled plasmas. A K_i–F_e–F_S simulation would use fluid descriptions of both the electron density (which largely controls the electric field) and the source rate. The kinetic ion model would give a detailed prediction of the ion distribution striking the walls, for instance, but the accuracy would be limited by the failings of the electron model. However, a F_i–K_e–K_S hybrid provides a relatively accurate description of the electrons. The fluid ion model is probably more reasonable than a fluid electron model and could also be used in an iterative scheme employing a kinetic ion model for part of the time.

8.3.6 Construction of a Hybrid Simulation

In the rest of this section we review in more detail some of the likely elements of a simple hybrid model that are appropriate for low-pressure plasma processing applications.

Ambipolar Diffusion

The ions in most plasmas we are studying undergo "ambipolar diffusion." The ion flux is driven primarily by the electric field, $\Gamma_i = n_i \mu_i \mathbf{E}$, but since $\mathbf{E} \simeq -\frac{k_B T_e}{e} \frac{1}{n_e} \nabla n_e$, $\Gamma_i \simeq -\frac{k_B T_e}{e} \mu_i \frac{n_i}{n_e} \nabla n_e$. When the plasma is quasineutral, $n_e \simeq n_i$, then $\Gamma_i \simeq -D_a \nabla n_i$, with $D_a \equiv \frac{k_B T_e}{e} \mu_i$. So the ion transport can be described as diffusive with diffusion coefficient D_a. If we know the ionization rate S then we can find the density from it by solving the diffusion equation. This is the simplest F_i module that is likely to be useful.

Finding the ionization rate is much more difficult than solving the diffusion equation. The simplest way to find the ionization rate, which is likely to work in a broad class of physical circumstances, is probably to use a Monte Carlo method.

Monte Carlo Calculation of the Ionization Rate

The density found in the ambipolar diffusion calculation, described above, is to be used to indicate where electrons can be launched in the (K_S) Monte Carlo calculation. If the density in a cell of area Δs is $n(\mathbf{r})$ then the number of Monte Carlo particles launched from there is $\alpha n(\mathbf{r}) \Delta s$. For now we assume α to be a constant. (If α varies with position, we will need to keep track of the "weight" of each Monte Carlo electron.) In addition, if secondary emission of electrons from the surfaces is important then the number of electrons emitted must be found from the ion flux to the surface (which is known from the ambipolar diffusion calculation) and the secondary emission coefficient γ_s. The electrons launched in the plasma volume are each (in principle) to be followed (and when they cause an ionization the time and place recorded) until they escape from the plasma, or until some maximum integration time has elapsed. The secondary electrons are also each followed in the same way.

There will frequently be situations where we need α to be a strong function of position. If the electrons in sparsely populated regions (including sparsely populated regions in velocity space) are responsible for most of the ionization, then we need to follow them and not waste time following the much more numerous electrons in high-density regions that will almost never ionize a neutral. In such a case we must keep track of how many real electrons are represented by each "simulation electron."

8.3.7 Feedback in a Plasma Simulation

The steps described above, and frequently other calculations, must be performed repeatedly until the simulation converges. For convergence to happen, there must be negative feedback in the simulation. If the density doubles between one iteration and the next (for example), there must be some physical mechanism included in the model that reduces the ionization rate per electron (or increases the loss rate per electron). If the ionization rate per electron did not go down, in our example, the density would double at every iteration and then the simulation would not converge.

A possible feedback mechanism that will reduce the ionization rate per electron (IRPE) occurs when the increased density decreases the electric field responsible for heating the electrons. A few examples of this are:

1) In a positive column a fixed voltage is applied to the plasma. The plasma is in series with a ballast resistor. When the plasma density goes up, the plasma current goes up, the voltage dropped across the ballast resistor goes up, and the plasma sees a lower applied voltage; this is one mechanism which reduces the IRPE. (In a positive column, this is not the only mechanism involved.)

2) In an inductive discharge, an increase in the plasma density causes a decrease in the skin depth, which is the characteristic distance that the rf fields penetrate into the plasma. Then a smaller fraction of the electrons are in the heating region and a lower IRPE results.

3) In a "capacitively coupled" discharge, the heating is sometimes caused by the moving sheaths. When the charge density ρ goes up, the sheath thickness x_s goes down. This is because for a uniform charge density ρ, with $E = 0$ at the plasma edge of the sheath, a voltage Φ is dropped in a distance x_s, where $\Phi = \frac{1}{2}(\rho/\varepsilon)x_s^2$. Since the sheath speed is of the order of $\omega_{rf}x_s$, which goes down, electrons "bouncing off" the moving sheath get a smaller kick. (If the electrons are thought of as being inside the sheath, then the sheath is smaller and thus the electrons spend less time there and there will be fewer of them inside.) This reduces the IRPE.

In summary, it should be expected that the heating mechanism of the electrons will have to be described carefully for use in the MC calculation. This mechanism will probably provide the feedback needed for the plasma to reach steady state, and it is vital that this mechanism be included properly in the simulation, if a self-consistent result is to be obtained.

8.3.8 Direct Numerical Calculation of the Distribution Function

A kinetic description of particles can be achieved by means of the distribution function. (Computation of the distribution was discussed by Rees [224], although Rees's method has not been used extensively. For examples of kinetic studies in plasma processing, see Refs. [225–233]. A general review of treatments of the kinetic equation is given in Ref. [234].) We shall now extend our earlier discussion of direct solution of the kinetic equation to allow us to find the distribution function f in general situations.

The distribution function f is the number of particles per unit volume and per unit of each of some appropriate velocity variables. The kinetic equation describes the rate of change of f with time and if there is no production or loss of particles it guarantees that the number of particles is conserved. The movement of particles in space and (ignoring collisions) their acceleration, which is a movement in the velocity variable, can be shown not to compress the distribution in "phase space." If the particles in some small region of position and velocity are observed, the region containing those particles may stretch in space, but if it does, it will shrink its extent in the velocity coordinates to keep the overall phase-space volume the same. The conservation of particles, combined with the incompressibility of f, means that as a particle moves around in space and velocity, the density f in the vicinity of the particle does not change (unless collisions or ionization change the number of particles in the volume). The density does not change because the volume in "phase space" of the region containing the particles does not change, nor does the number of particles in the volume.

If $\nabla \cdot \mathbf{v} = 0$, where $\nabla \cdot \mathbf{v} \equiv \frac{\partial v_x}{\partial x} + \cdots$ and $\nabla_{\mathbf{v}} \cdot \mathbf{a} = 0$ where $\nabla_{\mathbf{v}} \cdot \mathbf{a} \equiv \frac{\partial a_x}{\partial v_x} + \cdots$, then the flow in phase space is incompressible. Now $\nabla \cdot \mathbf{v} = 0$ because the components of $\mathbf{v}$, like the components of $\mathbf{x}$, are the independent variables in the space we are working with. $\nabla_{\mathbf{v}} \cdot \mathbf{a}$ can also be shown to be zero, for an acceleration $\mathbf{a}$ caused by the Lorentz force.

On the other hand, for conjugate momenta and coordinates (p_i, q_i) obeying Hamilton's equations, with Hamiltonian $\mathcal{H}$,

$$\dot{p}_i = -\frac{\partial \mathcal{H}}{\partial q_i} \tag{8.68}$$

and

$$\dot{q}_i = \frac{\partial \mathcal{H}}{\partial p_i}, \tag{8.69}$$

the equations of motion guarantee that

$$\frac{\partial \dot{p}_i}{\partial p_i} + \frac{\partial \dot{q}_i}{\partial q_i} = 0. \tag{8.70}$$

If the particles do not only move around but also undergo collisions and so forth, then the number in the volume can and will change. Processes, such as ionization, which add or subtract particles can change the distribution further. Collisions (except ionization events) do not add particles, but collisions do move particles "suddenly" from one velocity to another without changing the position of the particles in space. (A collision can be defined as a "sudden" interaction between particles. As such it is an irreversible process. What is considered "sudden," and consequently irreversible, depends on the observer.)

A method for studying the behavior of a collisionless plasma, which exploits the incompressibility of f, was developed by Berk and Roberts [235]. Their method was called the Water Bag method. The method was applied to one space dimension z and one velocity dimension v_z. The two variables (z, v_z) define the phase space in this case. Within this phase space, contours of constant f were drawn. The contours were labeled by a number of points at intervals all around the contour. The points and the contour defined by them move together as time progresses. This contour can then be followed as it moves around during some interval in time, by integrating over time the motion of the points that label the contour. By following each contour in this way the entire distribution can be evolved in time.

The region between each successive pair of contours is a Water Bag. The "density" f inside each Water Bag is approximated as being uniform throughout the bag. The method was used effectively to study kinetic plasma instabilities. The problem with the method is that the shape of the bags becomes complicated enough, over time, that the labels are too few to describe the bags. It is also unclear how well the labeling will work in a phase space with more variables.

8.3.9 The Convected Scheme

A technique called the "Convected Scheme" (CS), which uses some ideas similar to those behind the Water Bag model, will be described next. (A simple treatment was given in Chapter 4.) This new method has been applied in phase spaces ranging from (z, v_z) to a five-dimensional phase space with two physical-space and three velocity-space coordinates, which were all used as independent variables [67, 70, 82, 236–244]. Different choices of independent variables help to make different aspects of the calculation more accurate. The variables that are conserved by the collisionless motion are convenient for describing the collisionless motion. Speed and angle in velocity space are the best choice for accurately describing the redistribution of particles after a collision. For instance, in a problem where one space variable, z, is employed, collisions are more easily described in (v, μ) velocity variables, whereas ballistic motion is more accurately handled using $(v_z, v_\perp)$. There is a mismatch between the ballistic motion and the collisions, in this regard, which is a major part of the challenge in setting up an accurate simulation [82, 237, 240, 241, 244].

The same (z, v_z) phase space will be used as an example, but now each of these variables will be divided into intervals, of width Δz in z and Δv_z in v_z, creating "cells" of size $(\Delta z, \Delta v_z)$. Each such cell can be thought of at the start of a time step as being a Water Bag, since the goal of the method is to move the density in each cell in (nearly) the same way as the Water Bags are moved. The cell does not move; only its contents will be moved like a Water Bag. The main difference from the Water Bag method is that (in the simplest version of the CS) each of these Water Bags is followed for only one time step Δt before the density in the bag is split up. The density is then shared between the cells that the bag overlaps after moving for a time Δt. The CS is illustrated in Fig. 8.3.

At the start of a time step at time t each Water Bag coincides with a cell of the mesh; the bag is then moved during the time step Δt. At the end of Δt the bag overlaps several cells, which may include the initial cell. The initial density in the initial cell is the same density that is in the Water Bag. The fraction of the bag that overlaps each cell, at the end of the time step at time $t + \Delta t$, is the fraction of the density that is put in that cell.

In practice it is not feasible to move the water bags in exactly the manner described above, since the labeling of many points around the boundary of each bag would require a very large number of total points. In addition, the stage in the calculation where density is added back into the cells of the mesh must be handled carefully if serious errors are to be avoided. For each cell that is moved, a number of quantities must be conserved exactly. First the number of particles must be conserved; this is easily done if the initial cells and the particles leaving them are considered, one initial cell at a time. If the fractions of the particles from a given initial cell that are to be added to each final cell are known, the fractions can be normalized to sum to exactly one. Then the exact same number of particles that were in the initial cell will be returned to the mesh, at the end of the step, in some group of cells.

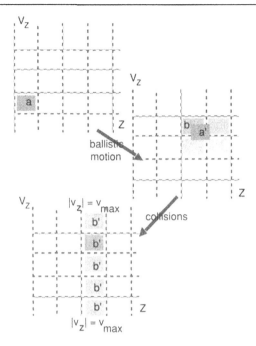

Fig. 8.3. Ballistic motion and collisions in the convected scheme are implemented sequentially. The density in cell a goes to the position of the moving cell a′ owing to the flow in position and velocity. Moving cell a′ overlaps several cells b on the mesh, and the density in a′ may be returned to the mesh, in the cells labeled b. First the spatial overlap of a′ with the final spatial cells is used to find how much of a′ is in each spatial cell. Then, within each spatial cell, the particles that are being put back in that spatial cell are divided between velocity cells. The mean kinetic energy/speed in each spatial cell is found from conservation of energy, assuming the potential energy is constant within each spatial cell. Particles will be shared between velocity cells (one spatial cell at at time) to get the correct average energy or speed within that spatial cell. Collisions take some of the density in a cell b and spread it out to the cells b′. The cells b′ are at the same spatial location as each other, because collisions do not change the particles' position. The range of v_z shown might correspond to the situation where the speed v is given by $v^2 = v_z^2 + v_\perp^2$; therefore v_z cannot exceed v.

For many purposes it is also essential to conserve the total energy of the particles in the initial cell, when those particles are put back into cells of the mesh. This can be achieved by considering, one at a time, each spatial cell in which the particles are being added back. In that spatial cell the potential energy is assumed to be constant throughout the cell. We sometimes say that our potential is a "staircase," because of this assumption. The staircase assumption is necessary to make the calculation of the kinetic energy unambiguous. Then the kinetic energy of the particles added to the spatial cell can be found from their initial total energy and from the potential energy of the final cell. It is not usually possible to ensure that all the particles that are put back in a spatial cell individually have the exact kinetic energy required. Instead, the particles usually have to be shared between two energy cells, in the correct proportions, so that their average kinetic energy has the appropriate value. In many cases, it is necessary to use a similar procedure to conserve additional quantities such as the angular momentum. (If particles have not left their initial spatial cell, conservation of energy means their velocity is unchanged. The electric field is found from the difference in electrostatic potentials at the cell centers, however, so the particles do feel a force. There is an inconsistency in the handling of the potential, which appears to be unavoidable, which leads to an inconsistency in the velocity. The way this is handled is addressed in the exercises at the end of this chapter.)

To determine the number of particles going to each final spatial cell one uses the spatial faces of the bag, which are moved as the bag moves. The front face of one cell and the back face of its neighbor coincide, and they must continue to coincide as the bags are moved or the density of particles will develop artificial holes or bumps. The fractional overlap of the bag with the spatial cells of the mesh determines the fraction of the particles going to each spatial cell. In other words, if at the end of a time step 20% of the spatial extent of the bag falls in a particular spatial cell of the mesh, then 20% of the particles in the bag are put into that spatial cell at the end of the time step.

The method described here enables one to find the fractions of each initial cell that go to each final cell. The expressions for the fractions are sometimes called "Green's functions" or "propagators." This technique for finding those fractions could almost be called an "Ephemeral Water Bag" method, to use Eastwood's terminology [204, 205], since the Water Bags each only last for Δt. Instead it was named a "Convected Scheme" (CS) because it describes the convection in a natural way. The CS does not actually duplicate the Water Bag model closely, even within a single time step. Further, the CS is used in multidimensional phase spaces in collisional plasmas as was mentioned above. The Water Bag model is nevertheless the closest relative of the CS.

Convected Scheme Using Long-Lived Moving Cells

A major challenge in solving the kinetic equation numerically is to minimize numerical errors that occur because of finite mesh size [82, 241]. The choice of an appropriate set of independent velocity-space variables can help: $(v_z, v_\perp)$ minimize errors in the collisionless motion, whereas (v, μ), where $\mu \equiv v_z/v$, are more accurate for collisions. An innovation that can increase the accuracy of the ballistic (collisionless) part of the CS involves allowing the moving cells to last for more than a single time step [244]. This technique might permit the independent variables to be chosen to focus on making the description of collisions more accurate.

As before, each of the moving cells starts out by coinciding with a fixed cell of the mesh. However, these moving cells are followed for many time steps, instead of for a single time step. Right behind each moving cell is another moving cell (Fig. 8.4). Every time a moving cell completely leaves the initial cell where it originated, a new moving cell fills the same initial cell. Particles that collide and are returned to the mesh in a given fixed cell are placed in whichever of the moving cells that originated in the fixed cell is currently closest to the center of the fixed cell. The moving cells are followed for up to (about) two collision times after they left the fixed cell. Since no particles are added to the moving cell after it leaves the fixed cell, then after two additional collision times, roughly 90% of the particles in

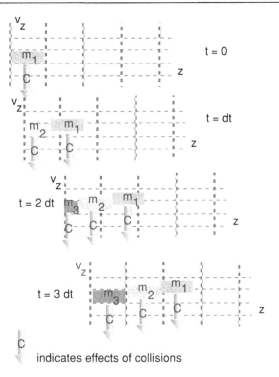

Fig. 8.4. Long-lived moving cells in the convected scheme reduce numerical diffusion, because those of the particles in the moving cells that do not have collisions are "mapped back" onto the fixed cells of the phase space mesh only after a "long" time has elapsed. m_1 is the first moving cell launched from a particular fixed cell on the mesh; m_2 is the second moving cell launched from that fixed cell; and so on. Cell m_1 will be mapped back onto the fixed mesh before m_2, and so forth. Collisions take some of the density out of the moving cells at every time step, however. This density must be returned to other cells corresponding to the correct velocity that the particles have after the collision.

the moving cell have already scattered. At this stage there is little to be gained by following the moving cell further, and so the rest of the particles in the moving cell are returned to the fixed mesh, as is done in the usual version of the CS.

The advantage of allowing the moving cells to last longer is that "numerical diffusion" is reduced by taking long time steps. The disadvantage of this is the need to keep track of several times as many cells as in the usual CS. The greatest technical difficulty in implementing the CS is dealing with cells that bounce. When using moving cells, it may be appropriate to map the contents of the moving cell onto the fixed cells, whenever part of the moving cell bounces, to avoid having to try to keep track of a cell whose parts are trying to go in different directions.

Collisions in the Convected Scheme

Collisions do not directly change the positions of particles. They change the particles' velocities, which will eventually lead to a change of position. The direct effect of the collision is thus to rearrange where particles are in velocity space, within the same spatial cell.

The CS is primarily designed to describe the evolution of the distribution functions of charged particles. The collisions that charged particles undergo are of several types, depending on the particles themselves and the neutral gas through which the particles are moving. The most important collisions the ions undergo in a partially ionized plasma are with the neutrals: elastic collisions and charge-exchange collisions. The main collisions the electrons have are elastic and inelastic collisions with neutrals, and in some discharges Coulomb collisions with other electrons.

Collisions with neutrals are straightforward to handle [67, 70, 240, 242] in the CS. In any given type of collision of a charged particle with a neutral and for any given initial kinetic energy of the charged particle, the probabilities of going to all the possible final kinetic energies can be calculated once at the start of the calculation and stored in a table for later use. The sequence of handling the physical processes within a time step Δt might then be:

i)   Move the particles in each cell of the mesh in phase space, according to their velocity and acceleration, with a time step Δt, as described in the previous sections.

ii) Find the fraction of the particles that collided in Δt. If ν is the collision frequency for any given type of collision, then the fraction that collide in that type of collision in the time Δt is $\nu \Delta t$, provided $\nu \Delta t$ is small. Redistribute the particles that collided, within the same spatial cell where they collided, to the energies indicated by the table, for that type of collision and for that initial kinetic energy. Repeat this for all types of collision. Go back to step i.

Coulomb collisions between electrons can be important in redistributing energy between the electrons. The electrons' momentum, but not their kinetic energy, is changed rapidly in elastic electron–neutral collisions. Coulomb collisions are unlikely to be frequent enough to have a similar effect on the electrons' momentum. However, Coulomb collisions can sometimes change a low-energy electron's energy as fast as the competing processes that affect low-energy electrons.

To show how the table is set up that describes to which energy particles go after they have a collision, two simplified examples will be given. The first will address charge-exchange collisions of ions, and the second will be an example of elastic collisions of electrons with neutrals.

1.   When an ion has a charge-exchange collision with a neutral, an electron is exchanged, turning the fast-moving ion into a fast-moving neutral. The slow neutral becomes a slow ion. The effect of the collision is to take a fast ion and replace it with a slow ion. If the old and new ions are of the same species, the result is as if the same ion had lost nearly all its energy. All ions having a charge-exchange are returned at the lowest energy (in a very simple treatment, or in a range of low energies) with a probability of one. The table for charge-exchange collisions, therefore, typically has an entry of unity for the probability of going to the lowest energy cell once the ion has had a collision, and this is the simplest way to return a particle that has had a charge-exchange collision, whatever the initial ion energy.

If the time step is very long, it will sometimes occur that ions that have just had a collision have enough time to leave the cell where they are placed after a collision, within the same time step in which the collision occurred. In other words, the ions placed in the cell by collisions or ionization during the first part of the time step can leave before the end of the time step. In reality ions are being added to the cell by collisions or ionization continuously and are flowing out of the cell. If we only add them at discrete intervals – once per time step – then if they flow a long way in a time step, gaps will appear between the packets of particles that we add. (Particles added to cells by the ballistic move do not suffer from this problem, because neighboring cells moving with the flow are all entrained in the same flow. However, if we do not move the cells accurately enough, gaps can still appear.) This must be handled by allowing for the movement of the ions, after the collision and during the same time step in which they collided. An analogous situation is described in the next section, on propagators for diffusion processes, when "source terms" are considered.

2. Elastic electron–neutral collisions cause the electron's energy to decrease very slightly. If one ignores the small loss of energy (which in practice is not done) the electrons must be returned at the same energy but with different velocities, that is, different directions of motion [82].

The problems at the end of this chapter go into more detail on the use of various versions of the CS.

8.3.10 Propagator for Diffusion

We conclude this chapter with an efficient numerical method for solution of diffusion equations. One way to solve a diffusion equation is to replace it with a finite-difference equivalent equation. Another method, which is often easier to use, uses a "Green's function" method, wherein if we know how density collected at a point will spread out in space as time passes, we can use this knowledge to calculate how a more realistic initial density will spread out.

To begin with, we shall consider a two-dimensional region labeled by coordinates (x, y). Each coordinate axis will be divided into small equal intervals, Δx and Δy respectively. A "cell" of area $\Delta x \Delta y$ will be treated as being in some circumstances small enough to be considered as a "point" in what follows.

In two dimensions, an initial density at time t' consisting of one particle at a point $\mathbf{r}'$ (i.e., a delta function density) gives rise to a density

$$n = \frac{1}{4\pi D(t - t')} \exp\{-(\mathbf{r} - \mathbf{r}')^2/4D(t - t')\} \tag{8.71}$$

at any point $\mathbf{r}$ at a later time t. This is a Gaussian shape, and the width of the Gaussian is of the order of the standard deviation, $\sigma \sim \sqrt{2D(t - t')}$. This will be used to represent the density coming from an initial cell with sides of length Δx and Δy. For this to seem like an initial point, the standard deviation σ must be much larger than Δx or Δy.

The number of particles in the area $\Delta x \Delta y$ is the initial density multiplied by $\Delta x \Delta y$. Suppose we label the center of each cell by a pair of integers (i, j) so that at the center $x = i\Delta x$ and $y = j\Delta y$. The density in a cell at time t will be written as $n(i, j, t)$. The density at the center of the initial cell, at time t', will be denoted $n(i', j', t')$, and the number of particles in the cell at time t' will be $n(i', j', t')\Delta x \Delta y$.

Some other cell will have its center at $x = i\Delta x$, $y = j\Delta y$. At some later time t the density in this other cell, due to the density which was in the initial cell at time t', is

$$n(i, j, t) = n(i', j', t') f(\delta i, \delta j, \Delta t), \qquad (8.72)$$

where

$$f(\delta i, \delta j, \Delta t) = \frac{\Delta s}{4\pi D(t - t')} \exp\left\{-\left[\frac{(i - i')^2 \Delta x^2 + (j - j')^2 \Delta y^2}{4D(t - t')}\right]\right\},$$

$$(8.73)$$

and $\Delta s = \Delta x \Delta y$. The quantity f depends only on the spacings $\delta i \equiv i - i'$ and $\delta j \equiv j - j'$ and on the time step $\Delta t \equiv t - t'$.

If Δx and Δy are very small, the function f given here is already (nearly) normalized so that particles are conserved. In practice, because Δx and Δy are finite, it is necessary to make sure particles are conserved by supposing (in a test case) that there is one particle in the initial cell (so $n(i', j', t')\Delta s = 1$), and by insisting that exactly one particle is put back. Now the density in the final cell must be multiplied by Δs_{final} (which equals Δs in this case) to give the number of particles in the cell. Therefore

$$\sum_{i,j} \Delta s\, n(i', j', t') f(\delta i, \delta j, \Delta t) = 1. \qquad (8.74)$$

The sum over i and j is over all the final cells (in practice, out to plus and minus a few σ in x and y). If $\Delta s_{\text{initial}} = \Delta s_{\text{final}}$ then $n(i', j', t')\Delta s = 1$. However, $\sum_{i,j} f(\delta i, \delta j, \Delta t)$ will only be approximately equal to one because of the finite cell size. To correct for this, we redefine $f(\delta i, \delta j, \Delta t)$ to be the original f multiplied by a constant A. Then we require that $\sum_{i,j} f(\delta i, \delta j, \Delta t) = 1$, and we find the value of A from this equation.

Source Terms

New particles are produced at a rate S per unit area per second, and so in a cell of area Δs in a time step Δt there will be $S\Delta s \Delta t$ particles added. This is not a reference to particles diffusing into the cell, but to particles "born" in the cell. It is sometimes reasonably accurate to simply add this number to the number of particles in the cell at the end of the time step. However, it is more accurate to allow the particles that are added (say) halfway through the time step to diffuse for the remaining half of the time step.

Suppose we divided the time step into r equal intervals. In the first interval of $\Delta t/r$ an extra number of particles equal to $S\Delta s \Delta t/r$ is added. These are then allowed to diffuse for the rest of the time step, or for a time $\Delta t_1 = \frac{(r-1)}{r}\Delta t$. They can then be added to the initial cell and the neighboring cells at the end of the time step. Similarly, in the second time interval, the same number of particles is added. However, they have less time, $\Delta t_2 = \frac{(r-2)}{r}\Delta t$, to diffuse until the end of the time step. The particles added in the nth interval in cell (i', j') make a contribution at the end of the time step to the density in cell (i, j) equal to

$$S\frac{\Delta t}{r}f(\delta i, \delta j, \Delta t_n). \tag{8.75}$$

The particles added during all r substeps contribute a density in the cell (i, j) equal to

$$S\Delta t\, g(\delta i, \delta j); \tag{8.76}$$

where

$$g(\delta i, \delta j) \equiv \frac{1}{r}\sum_{n=1}^{r} f(\delta i, \delta j, \Delta t_n). \tag{8.77}$$

As usual, each of the fs for each value of Δt_n must be normalized.

Because $f(\delta i, \delta j, \Delta t)$ and $g(\delta i, \delta j)$ are used repeatedly, it is useful to generate them once and store them.

Boundary Conditions

The Gaussian shape of the density that came from an initial point is appropriate in a region with no boundaries. Suppose we have a plane boundary that either a) reflects all the particles reaching it or b) absorbs all the particles. These boundaries can be handled by the method of images. For each cell in front of the boundary, a cell of the same size behind the boundary must be included, at an equal perpendicular distance from the boundary. The density has the same magnitude, but has opposite sign for an absorbing boundary and the same sign for a reflecting boundary.

Partially absorbing boundaries can be handled with images whose magnitudes are smaller than those of the original cells.

Exercises

1. Repeat the calculation of the density that was done in Section 8.1.2, but this time use a finite-difference scheme to solve the diffusion equation. Perform the calculation in Cartesian coordinates and in cylindrical coordinates (see [29]). Comment on the difficulty of doing the analytic calculation in cylindrical coordinates as opposed to the numerical calculation.

2. A particle has a chance of roughly dx/λ of suffering a collision when it travels a distance dx, provided $dx \ll \lambda$. Let the step size Δx be such that $\Delta x/\lambda$ is 0.3 in total for collisions of any kind, while $\Delta x/\lambda = 0.1$ for ionization events. What is $\Delta x/\lambda$ for all processes except ionization? This question involves using a telephone directory as a "random number generator" and performing a Monte Carlo simulation, in one dimension, by hand. Use the last digit of each of a

column of numbers as the random number r. A collision will be said to have occurred for $r \leq n$. What is n? What is n for an ionization event? Use the second to last digit to determine if the particle changes direction, if the last digit indicates that a collision occurred.

Sketch the particle's progress, in one dimension, horizontally. Draw its steps as being of equal length, Δx, but move the lines slightly downward each time the particle changes direction. Choose the size of the step so that many steps fit on the page. Mark ionization events with an x. Continue until the particle has had five ionization events. Repeat this process for three or four particles.

How many steps do you expect to take per particle? How far away from the start (in terms of λ) do you expect to end up?

3. Students forget to bring their text to class three times out of ten. What is the probability that all 50 students in a class will a) bring their text, b) forget their text, on any given day?

    What is the probability that c) exactly half the class will bring their text, d) at least half the class will bring their text?

4. A plasma is contained in a cylindrical chamber, with radius $a = 30$ cm and height $h = 25$ cm. The electron temperature T_e corresponds to about 3 eV. The electron and ion number densities inside the plasma have peak values of about 10^{12} cm^{-3}.

    The ion density in the sheath is about $f_s \lesssim 0.1$ times the peak density. Estimate

    a) the Bohm velocity, for an Argon plasma.
    b) the sheath potential, assuming there are no large bias voltages applied.
    c) the sheath thickness.
    d) the flux of ions to the (entire) wall.
    e) the ion confinement time.
    f) the ion energy flux to the entire wall. (Neglect collisions in the sheath.)
    g) the electron energy flux to the wall.
    h) the power required for ionization of neutral atoms.
    i) how much power is being used in making "hot" neutrals by means of charge exchange, if a typical ion has an energy of 0.25 T_e, and the mean free path for charge-exchange collisions is $\lambda_i^{cx} \simeq 5$ cm.
    j) how many times as much power goes into excitation of neutrals as goes into ionization.
    k) the total power input to the plasma, neglecting any ionization by metastables colliding with each other.

5. See Fig. 8.5. This example attempts to illustrate how to solve a problem in kinetic theory by means of an analogy. The analogy is that we have built a machine to pump water, by pumping it into paper bags that are picked up and accelerated to the east end of the room. The paper bags are started in a rectangular pattern. The paper bags at the south end of the room are not moving when they are set down. The bags further to the north, however, are moving from the outset. When the bags are picked up, they are lifted over the area where the bags are started, and as they go faster east, they are also moved north, so that

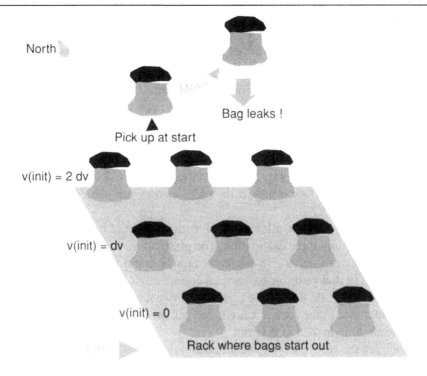

North

Pick up at start

Bag leaks !

v(init) = 2 dv

v(init) = dv

v(init) = 0

Rack where bags start out

Fig. 8.5. The bursting paper bag analogy for kinetic simulation. The figure shows paper bags representing the moving cells that carry the density around in position and velocity. The bags leak; the leaking represents collisions. They also burst after a period of time – this represents the process in which we stop using a particular moving cell to describe the fate of the particles and put them into a new set of cells that will describe where they are.

the bags starting underneath are moving at the same speed as the bags being carried above them.

The paper bags leak, so that water spilling out of them falls on the floor, drains due south, to the south of the room, and is pumped into the stationary bag that is as far to the east as where the water leaked out.

The bags also burst after some amount of time. When they burst, their contents are caught in the bags being set down in the rack beneath them, as far to the east and north as the bag that burst. Because the bag in the rack is as far north as the bag that burst, each is moving at the same speed. When the bag bursts, it will usually overlap more than one bag in the rack, and so its contents have to be shared between several bags in the rack.

A new set of bags are set down and started in motion every second. Ten percent of each bag leaks out in a second. We have bags of two strengths. One type of bag bursts after one second. Suppose that after one second the bag is 30% of the way over the next position to the east, this position referring to the position of a bag in the rack where the bags are started. It is also 20% of the way north, over the next position in the rack.

Then the bag that is moving is (0.3×0.8) over the position one step east and zero steps north; it is (0.3×0.2) over the position one step east and one step north; it is (0.7×0.8) over the original position; and it is (0.7×0.2) over the position zero steps east and one step north. The water is shared between new bags, in these proportions, when the moving bag bursts. These numbers are for illustration only. The actual fractional overlaps depend on the speed of the moving bag, which depends on how far north it is.

The fractional distance the bags move east per second is $0.1 \, N$, and the distance they move north is 0.2, where N is an integer that labels how far north the bags started. N runs from zero to twenty. There are 15 bags in the west–east direction. One unit of water is added in the most westerly and most southerly bag every second. Set up a simulation to calculate the amount of water being put into each bag in the rack, as they are started, at one second intervals.

Repeat the calculation, allowing for a stronger type of bag; this might be the plastic bag version of the calculation. The plastic bags do not burst until 20 seconds after they are started. In this case, new bags are not put in the rack until the last bag in that position has moved completely out of its original position in the rack. It can do this by moving north or east (or both). Water falling into a particular slot in the rack is caught and put into whichever moving bag was most recently launched in that slot.

Compare the results of the two calculations.

6. Write a computer program to solve the time-dependent, electron kinetic equation, using the independent velocity variables $(v_z, v_\perp)$, which are to be represented by discrete values on a mesh. For now, there is no independent spatial variable. Assuming there is an electric field in the z direction, the z component of the field being E_z, then in a time step Δt each electron has its v_z change by an amount

$$\Delta v_z = -\frac{eE_z}{m} \Delta t. \qquad (8.78)$$

$v_\perp$ is not changed, directly, by E_z.

In the time Δt, a fraction p^{el} of the particles, $p^{\mathrm{el}} \simeq v_e^{\mathrm{el}} \Delta t$, in each cell undergo elastic collisions (provided $p^{\mathrm{el}} \ll 1$). Here v_e^{el} is the electron elastic collision frequency, which in general depends on $v_e = \sqrt{v_z^2 + v_\perp^2}$. Similarly, a fraction p^{inel} of the particles, with $p^{\mathrm{inel}} \simeq v_e^{\mathrm{inel}} \Delta t$, undergo inelastic collisions, provided $p^{\mathrm{inel}} \ll 1$. The frequency v_e^{inel} is zero below a threshold in v_e corresponding to the threshold kinetic energy $\varepsilon_{\mathrm{th}}$ for the inelastic process. Assume electrons having elastic scatters are redistributed at the same v_e as before the collision, but isotropically, with equal numbers in equal intervals of v_z, from $v_z = -v_e$ to $v_z = v_e$. $v_\perp$ is then chosen to give the correct v_e; $v_\perp^2 = v_e^2 - v_z^2$. Since the mesh points are at $(v_z = i \Delta v_z, \; v_\perp = j \Delta v_\perp)$, it will not be possible to find the exact value of $v_\perp$ required at any given v_z. Because of this, particle density is shared for each v_z between the two values of $v_\perp$ that are on the mesh, on either side of the true $v_\perp$. The density is shared in such a way that the average kinetic energy of the particles in the two $v_\perp$-cells is correct. Particles having inelastic collisions are also redistributed isotropically, but at a lower kinetic energy, so that the new electron speed is

$$v_e = \sqrt{v_e^2 \text{ (old)} - 2\varepsilon_{\text{th}}/m}. \tag{8.79}$$

The calculation of the effects of the collisions can be made relatively efficient by calculating the list of final cells, for each type of collision and for each initial velocity cell. The fraction of the scattered particles that go to each final cell can be calculated and stored for later use, since the fractions are the same, every time a collision of a given type occurs.

For an initial distribution that is a monoenergetic beam, with kinetic energy ε_i, and $v_\perp = 0$ so $\frac{1}{2}mv_z^2 = \varepsilon_i$, calculate the time evolution of the beam. First use $\varepsilon_i < \varepsilon_{\text{th}}$. Use a constant elastic collision frequency v_e^{el}. Choose the time step so $v_e^{\text{el}} \Delta t \lesssim 0.2$. Comment on how long it takes for the distribution to become isotropic. If the distribution spreads out significantly in energy, this is due to numerical diffusion. Try to minimize this by using a fine mesh. Repeat the exercise, using $\varepsilon_i > \varepsilon_{\text{th}}$, and include inelastic collisions, with $v_e^{\text{inel}} \ll v_e^{\text{el}}$.

Add an electric field in the z direction, $\mathbf{E} = E_z \mathbf{a}_z$. Repeat the exercise, for $\varepsilon_i < \varepsilon_{\text{th}}$.

7. This problem involves numerically calculating the ion distribution function in a one-dimensional plasma, using different versions of the convected scheme (CS). The phase space used is (z, v_z), and so the distribution function is $f \equiv f(z, v_z)$. First, consider the collisionless motion. The various versions of the CS all require a phase space mesh. Set up a mesh, with cells of width Δz in space, and with mesh points at discrete points in velocity v_z. For convenience, the velocity mesh points should also be uniformly spaced. (Neither part of the mesh needs to be uniform in general, – see Ref. [241] – but for now we choose to make it so.) $v_z = (i - i_0)\Delta v_z$ in cell i, and i_0 is an integer chosen to allow v_z to run from the most negative value selected to the most positive. This defines the "fixed mesh".

After each time step, the contents of each cell are moved. The shape of the contents in phase space and how far they move determine where they are put back (although they may be put back after one time step or after many time steps). The contents of a cell on the fixed mesh may move for several time steps; thus each set of cell contents, which began in some fixed cell, will be referred to as a "moving cell". At some point, each moving cell is returned to cells of the fixed mesh.

Each moving cell began at the location of a fixed cell, with a single, discrete value of v_z and a range of z values.

One way to update the location of the moving cell is to integrate the equations of motion of the front and back faces, for however many time steps we need. v_z is typically found in one of two ways. The first is from conservation of energy, so that

$$\frac{1}{2}mv_z^{\varepsilon^2} = \frac{1}{2}mv_{zi}^2 + e\Delta\Phi, \tag{8.80}$$

where v_{zi} is the initial velocity of the moving cell and $\Delta\Phi$ is the change in electrostatic potential, on moving between the initial location and the current location. The potential Φ is treated as being a staircase, that is, Φ is constant within each spatial cell. (Note that, if a particle stays in the same spatial cell, v_z^ε is unchanged. In this case, v_z^t, defined below, may be useful, especially if v_{zi} is small.) The second way integrates the equation of motion for v_z over time, giving

$$v_z^t = v_{zi} + \frac{eE_z}{m}t, \qquad (8.81)$$

where t is the elapsed time since the velocity was v_{zi}. We frequently use v_z^t for moving the front and back (in space) faces of the moving cell. We usually use v_z^ε to decide the velocity with which particles are returned to the fixed mesh. (The time used to find v_z^t can be chosen to be the larger of the actual time step, and a mean "residence-time" in the cell. See Ref. [241].)

If the moving cell overlaps, in space, with several spatial cells of the fixed mesh at the time when the return is done, then the fraction of the density put in each range of Δz of the fixed mesh is proportional to the overlap. In each spatial cell (or, equivalently, each range of Δz) of the fixed mesh, v_z^ε is found, as usual, from conservation of energy. Since v_z^ε found this way will not, in general, correspond to a single value of v_z that occurs on the fixed v_z mesh, the density is shared between values that are on the fixed v_z mesh and that are immediately above and below the actual v_z^ε found from conservation of energy. The proportions going to each v_z on the mesh are chosen to conserve velocity or energy. If v_z^ε is the true v_z, and the discrete values chosen are $v_z^{(i)}$ and $v_z^{(i+1)}$, then to conserve energy, if $f^{(i)}$ is the fraction placed at velocity $v_z^{(i)}$,

$$v_z^{\varepsilon^2} = f^{(i)}v_z^{(i)^2} + f^{(i+1)}v_z^{(i+1)^2}, \qquad (8.82)$$

where $f^{(i+1)} = 1 - f^{(i)}$. (We assume Φ is monotonic, and none of the particles bounced, for now.)

Now we introduce the next version of the CS we shall use, a long-lived moving cell method. Move the moving cells, first for as long as it takes for them to leave their initial cell, and, when they do leave their initial cell, start a new moving cell. When a new moving cell is launched, return the oldest moving cell associated with that fixed cell, placing its contents in cells on the fixed mesh. From there, put the contents (particle density) into the newest (hence closest) moving cell associated with that fixed cell. In this way, keep the same number, N_{mc}, of moving cells, for each fixed cell. The choice of N_{mc} will be discussed later. For now, let $N_{mc} = 4$. The particles may leave the initial cell because the back of the moving cell has moved a distance Δz in space. They may also leave in some versions of the CS, because v_z^t has increased, in the same spatial cell, to equal the next v_z on the discrete mesh. For now, we assume they leave the cell by moving in z rather than in v_z. However, if v_z does not change, some moving cells will not move properly, especially those with v_z near zero. For this reason, the moving cell has a "true" velocity v_z^ε, but it is also given a second value of v_z equal to

$$v_z^t = v_{zi} + \frac{eE_z}{m}t, \qquad (8.83)$$

where t is the time since its velocity was v_{zi}. v_z^t is used to integrate the motion of the cell faces in space. If v_z^t exceeds the next discrete value on the mesh, a new moving cell may be launched.

Before attempting this technique with a general "ballistic mover", it may be helpful to use a simplified CS.

The most straightforward way to update the positions of the moving cells is to assume they have an initial length, Δz, and an initial set of coordinates at the center of the moving cell, (z^{in}, v_{zi}). The length is then assumed to be constant, and the coordinates obey the equations of motion

$$v_z^t = v_{zi} + \frac{eE_z}{m}\Delta t,$$

$$z = z_i + \frac{1}{2}(v_z^t + v_{zi})\Delta t. \tag{8.84}$$

Suppose E_z is constant. Then v_z changes by the same amount in each time step, for every value of v_{zi}. Suppose the mesh spacing in v_z is Δv, and

$$\frac{eE_z}{m}\Delta t = \delta v, \tag{8.85}$$

where $\delta v = 0.3\,\Delta v$. Then

$$v_z^t = v_{zi} + 0.3\,\Delta v \tag{8.86}$$

and

$$z = z_i + v_{zi}\,\Delta t + 0.15\,\Delta v\,\Delta t. \tag{8.87}$$

In addition, for the collision process (see below) suppose $\nu\Delta t = 0.1$. Choose $\nu t_{mc} \geq 2$, where t_{mc} is the lifetime of the moving cell, so that $t_{mc} \geq 20\,\Delta t$. At each time step, 10% of the particles in any cell must scatter. Instead of forcing N_{mc}, the number of moving cells, to be an integer, follow each moving cell for 20 time steps, and then return it to the mesh.

At the start of the fourth time step after it was launched, the velocity of a moving cell (if moving cells last more than Δt) has increased by $1.2\,\Delta v$. If it has already moved in space by a distance $\geq \Delta z$, then a new moving cell should have been launched. Otherwise, a new moving cell must be launched after four time steps. Since the distance traveled depends on v_{zi}, then so does the frequency of launching moving cells.

Compare the distributions obtained in the method where each moving cell lasts Δt and the long-lived moving cell method. The second method should give rise to reduced numerical diffusion.

Finally, add collisions in each case. This is done by choosing Δt so that $\nu\Delta t \ll 1$, where ν is the collision frequency. Remove a fraction of the contents of each moving cell, the fraction being $\nu\Delta t$, at every time step. Assume the collisions reduce the velocity to zero. Place the particles that are scattered in the fixed cell at $v_z = 0$ and at the same spatial location as they scattered. In the second version of the CS, move the scattered particles, which are put into a fixed cell, from that fixed cell, into the nearest moving cell. This was described above for return of particles onto the fixed mesh and then onto moving cells. The number of moving cells N_{mc} is chosen so that each moving cell exists for a time t_{mc}, such that $\nu t_{mc} \gg 1$. This ensures that most of the particles in a moving cell are scattered out of the cell, before the entire moving cell is returned on the fixed mesh. (In practice, $\nu t_{mc} \gtrsim 2$ is probably sufficient.)

For what range of parameters is it useful to use the second version of the CS to reduce numerical diffusion?

9

Going Further

This book has suggested methods to begin to analyze the plasmas used for processing semiconductors, as well as the neutral chemistry used with the plasma, and the reactors used to create and contain the plasma. The material in this book should provide a solid base for understanding current research issues in this field and access to the relevant literature. To conclude this work, the appropriate way to describe these plasma systems will be considered, with emphasis on the tools needed to fully understand the plasma and to design a reactor.

Unlike most textbooks, there was no suggestion that we were giving a complete treatment of the problem. When textbooks give this impression, it is almost always misleading. This is particularly unfortunate, because it tends to discourage the reader from thinking creatively and independently about real problems, which cannot be handled in an oversimplified way.

To the extent that it is possible to formulate a general strategy for handling complex problems (such as processing plasmas), the strategy suggested is: Use the known experimental results and physical knowledge to set up simple models. Test the models by means of experiments. Refine the models and test them again. Repeat as often as necessary/possible.

It might seem that this process is the "hypothesis testing" that is supposed by philosophers of science to take place in all scientific endeavors. It is not the same, however. The sort of hypothesis they usually envisage is a single, straightforward idea. Whether or not hypotheses are ever arrived at and tested in the clear way that they imagine, the process of model building for a complex physical system is much more complicated. First, the model we need to construct involves many hypotheses. That is, there are many, important physical processes involved, all of which must be recognized, so that they can be included in the overall model. Second, we expect the hypotheses to evolve, during the process of experimental testing. This is in contrast to attempting to prove or disprove a single hypothesis.

The physical models that were stressed most in the previous chapters involved diffusion, random walks, and simple electrostatics. Circuit models were introduced, but they are very limited in their usefulness and should be thought of as being an analogy for the plasma behavior, rather than as providing a sound model. Circuit models should not be employed beyond the first, very simple, stage of model building.

More profound general arguments, based for instance on the energy balance in the plasma, can be used at the next level of model building. For example, in the case of energy balance one could establish bounds on particle production rates. Such arguments will probably not be capable of allowing determination of the exact state of the system. This is because details of the reaction rates, cross sections, and so on would not be a part of the type of general argument envisaged. It should also be possible to establish the nature of the "feedback" that is present in the combination of the plasma and reactor, early in the process of model building. It is likely that reasoning based on energy balance, possibly in combination with a calculation of the electromagnetic field in and around the plasma, will shed light on the feedback. This feedback is negative – it tends to drive the plasma density toward a stable value. If, for instance, the density grows above this stable density, the heating efficiency may be decreased so as to bring the density back down.

Approximate calculations, based on simple transport and electromagnetic considerations, can be used at this stage of the analysis. The very first transport analysis might involve estimates of particle and energy confinement times. The next level of transport analysis might require a calculation of a particle density, employing simplifying assumptions about the transport, the ionization rate, and so on.

To go beyond analytic estimates, it is necessary to employ numerical calculations. Even when quite complex computational models are used, it is difficult to set up a complete and accurate description of a plasma. For a discussion of the pitfalls of modeling in general, see Ref. [29].

9.1 Goals of Process Modeling

Having outlined a general strategy of model development, and before discussing modeling of a rather detailed nature, it is appropriate to assess the goals of modeling. The question to be answered first is: Do we want simple descriptions of the physical systems we study, or do we want very detailed predictions of behavior? Often the answer will depend on whether we are interested in basic science issues or in building an optimal reactor. In other words, are we concerned with basic science or with engineering? Ideally we should probably consider both. If we try to design a reactor (or set up a detailed model) without understanding the underlying mechanisms that control the behavior, we are likely to make mistakes. However, the motivation for understanding these systems at a fundamental level is to make them work better in reality. These arguments are intended to support the claim that we should be concerned with the science and the engineering issues.

As we have discussed elsewhere [29], the interaction of simple physical models and large-scale computational models provides another basis for comparison, like comparing either type of model with experiment, for iteratively improving the models. This may be particularly relevant, given the way large-scale models are sometimes set up and employed. It is not uncommon for models to be used, that are known or suspected to be inaccurate or to leave out important effects. If such a model is compared to experiment and it appears to work (in some sense, in some range of parameters), this is probably not a convincing test of the computational model. The simpler physical models should help to demonstrate when agreement with experiment is only coincidental and to prevent costly mistakes caused by the use of the computational model in inappropriate circumstances. Put more simply, if physical understanding says a computational model is flawed, then "successful" comparison to experiment does not fix the flaw – unless it shows us that our physical understanding was wrong. Faulty models are often set up because they are "better than nothing." This is a dangerous argument, but what is worse is attempting to justify a model that is known to be wrong on the basis of a few comparisons to experiment.

There is a tacit assumption in this discussion. In claiming that we need "simple" analytic descriptions of the process, and that we also need detailed computational descriptions, we are also implicitly stating that an analytic description of the process will not be complete. It seems clear, in practice, that processing reactors are complex in some sense. This complexity means that it is difficult to imagine what a simple analytic specification of the complete state of the reactor would be like. (Calculating this analytic result would presumably be even more difficult.) If it is true that systems such as this are complex (and it should not be assumed that they necessarily are) then there are serious consequences of this complexity. We are taught science in a fashion that implies that, if we were clever enough, we could always find simple solutions to scientific problems. It may be that much more effort needs to be put into finding ways to understand complex systems.

9.2 Detailed Description of Processing

The objective of our attempts to understand and to describe processing plasmas is to be able to effectively design and use improved reactors. The detailed description of the plasma that we seek is defined to be one that provides enough information so that we can deduce and specify how to set up a process that functions as well as possible, given the appropriate constraints.

There are several ways to obtain the detailed description needed. All of them require a great deal of effort. At one extreme, we could employ a series of experiments in some region of parameter space, to experimentally construct a "response surface" [245, 246] for each dependent variable of interest. A response surface is a fitted curve that interpolates the values of the dependent variable between experimental data points. This approach has advantages (such as its close contact with reality) and disadvantages (including that we may not know everything we would like to know about the independent variables controlling the response, and so our fitted surface may attribute some variation to the wrong cause; and that the surface is likely to have rapid variation, limiting its validity to regions that were sampled thoroughly, using many data points.)

The use of "neural nets" is in a way similar to construction of response surfaces. Both involve "fitting" to data. Neither directly provides an explanation of the data, although response surfaces are preferred by some because they can be used in finding an explanation. Nevertheless, neural nets provide one of the most highly developed detailed descriptions of processing. (This is a paradox, since in one sense they provide no description at all.)

Another approach, which is in a sense the opposite to experimental construction of a response surface, involves setting up a detailed computational description of the process. One advantage of the computational model is that the effects of the independent variables are clear – provided the model is correct and complete. In this context, "complete" means that it appropriately includes all the important effects. A disadvantage is that the model may not necessarily describe reality, unless it has been shown to do so by means of rigorous testing.

9.3 Detailed Computations

Computational reactor models provide one natural means to generate a description of a plasma reactor, which should be suitable for use in reactor design. One of the advantages of computational models is that they can handle many different physical subsystems in a global model of an overall physical system. Computational models of a subsection of the entire system are also extremely useful, especially if there is incomplete information as to the fundamental processes taking place in some of the subsystems which are not included in the model, and at the same time, if there are experimental measurements of the state of the subsystems that are not included in the model. In other words, if we know all the cross sections, and other fundamental physical data, we can set up a detailed physical model of all the subsystems. This is a time-consuming process, not only because gathering the data is difficult, but also because setting up and testing such a large-scale model is laborious. If we do not know the fundamental data, we might use a more limited model, in conjunction with measurements of the reactor parameters to make up for missing information. For example, to describe etching in a trench, we might measure the state of the main plasma, instead of simulating the main plasma from first principles. In either case, the trench model is furnished with information about the main plasma. The measured information should have the advantage of realism – it establishes contact with reality, provided the experiments are accurate and generate sufficient data. If we can compute the information about the main plasma, on the other hand, this allows us to explore the optimization of the main plasma. The computational model provides for extrapolation, whereas measured reactor parameters do not.

It appears that only a detailed computational model allows one to generalize sufficiently to provide an adequate description of a processing plasma, which will permit a global design optimization. All of the other techniques that have been considered so far, including analytic models, fitting of experimental data, and experiments themselves, are essential devices for testing such a model.

At present, plasma reactors are designed mostly by trial and error. It seems likely that this situation could change very rapidly. A computational model that operates based entirely on first-principles information is probably not realizable in the very short term, owing to a lack of basic data. Some input to the model of experimental data, on conditions in processing reactors, will be necessary for some time to come. The appropriate use of measured data to ensure contact with reality when fundamental data are absent, in conjunction with large-scale models, may provide an opportunity for progress (and for creativity) in the shorter term.

Glossary

Plasmas are capable of a truly enormous range of behaviors. Only a tiny fraction of these are likely to be important for materials processing applications. The majority of the theory introduced in the discussion of plasma reactors can be explained in terms of basic electromagnetism and the concepts of random walks. There is, however, a need for a separate section describing plasma theory that is devoted to explaining the terminology used, since that terminology tends to obscure the subject. This section will consequently provide a limited glossary of technical terms, emphasizing those terms that are not described elsewhere. The comments given here are not necessarily intended as technically exact definitions but rather as descriptive accounts.

Actinometry. Measurement of the gas phase densities of chemical species, by introduction of a known, small amount of a gas such as argon, which allows a calibration of the line emission intensity of the gas being measured.

Adiabatic Invariant. When a particle's motion is nearly periodic, an adiabatic invariant exists that is associated with the near-periodicity [247]; see also the article by R. J. Hastie in Ref. [34]. The adiabatic invariant is approximately conserved by the motion. The adiabatic invariant is the "action," or the area in phase space enclosed by the orbit of the particle (Fig. G.1). For particles gyrating (nearly periodically) around a (slightly) nonuniform magnetic field, the "magnetic moment" $\mu \equiv mv_\perp^2 / 2B$ is an adiabatic invariant. (For a uniform field, μ is a true invariant.)

Adsorption. Attachment of a molecule to a surface.

Alpha Regime. Used to describe operation of a discharge in a fashion in which ionization at each point is due to local heating of electrons by the electric field at that point.

Alumina. Aluminum oxide. Used as an insulating layer on the interior of many plasma reactors.

Anisotropic. See Isotropic.

Aspect Ratio. The ratio of two dimensions of an object. The aspect ratio of a rectangular trench is its depth divided by its width.

Asymptotic Analysis. Refers to the use of expansions in small parameters to solve equations, and so forth. Asymptotic series have the property that they may appear to converge, and the first few terms may even provide a useful result, but the series may then diverge as higher terms are included. See Ref. [53].

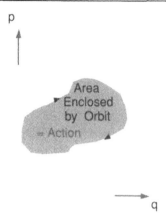

p

q

Fig. G.1. Illustration of an adiabatic invariant, which is equal to the area in phase space (p, q) enclosed by the (nearly periodic) orbit of the particle. q is a (spatial) coordinate and p is its associated momentum (or velocity).

Averaging.   See Method of Averaging.

Ballast Resistor.   A resistor placed in series with a dc discharge to provide negative feedback in the circuit. When the current goes up, more of the applied voltage is dropped across the ballast and so the voltage across the discharge goes down.

BGK Collision Operator.   See Krook Collision Operator.

Bohm Velocity.   The velocity v_B that ions are believed to need to pick up on average as they cross the "presheath" and before they enter the sheath [44]. v_B is approximately $(k_B T_e/m_i)^{1/2}$. The ion acquires a kinetic energy about equal to the mean electron energy, since the electrostatic potential in the plasma varies by about $k_B T_e/e$. If the ions' velocity were not $\geq v_B$ when they entered the sheath, the ion continuity equation and Poisson's equation would not be expected to have a reasonable solution in the sheath.

Boltzmann Equation.   The "kinetic equation" describing the time evolution of the distribution function. The kinetic equation can be thought of as a conservation equation in "phase space." The term Boltzmann equation usually refers to a version of the kinetic equation for a species a with a particular type of "collision term" that is regarded as appropriate for collisions between the species a in question and neutrals [34].

Brownian Motion.   The random walk of particles that leads to their diffusion. For motion in one dimension, with step size λ, the overall distance X traveled in N steps can be thought of as the result of N independent random events. In other words, each step is one of the independent random events. The distance traveled in one step has a mean value of zero and a variance of λ^2, so according to the Central Limit Theorem, the overall distance X traveled in N steps has a Gaussian distribution with zero mean and variance $\lambda^2 N$. If the frequency of taking steps is ν then after time t the number of steps is $N = \nu t$ and the variance is $\lambda^2 \nu t$. The probability density function is then $f_X(x) = [\lambda\sqrt{2\pi \nu t}]^{-1} \exp(-x^2/2\lambda^2 \nu t)$.

Bump-On-Tail.   When the high energy part of the distribution function, or tail, is not monotonically decreasing it may be said to have a bump on the tail. The bump on the tail may lead to an instability [43].

Capacitive Discharge.   A discharge where the power is supplied primarily by applying an oscillating voltage to conducting boundaries in contact with the plasma.

Cathode Fall.   A relatively wide region of strong electric fields adjacent to the cathode in a dc discharge, where electrons emitted by the cathode can "avalanche" to provide the ionization that sustains the discharge. The cathode fall usually has little visible light emission since the electron density there is low.

Cathodic Arc.   Arc formed on a cathode that is used to produce a "spray" of ions of the cathode material. Cathode material also tends to form molten droplets.

Central Limit Theorem.   Refers to a sum S of n independent random variables that have the same mean μ and variance σ^2, but where each random variable can have *any* distribution with a finite μ and σ^2. As n becomes large, the Cumulative Distribution Function of S approaches that of a Gaussian random variable. The mean value of S is $n\mu$ and the variance of S is $n\sigma^2$, and S is distributed so that $Z = (S - n\mu)/\sqrt{n}\sigma$ has a Gaussian distribution with zero mean and unit variance.

Charge-Exchange Collision.   A collision in which an ion loses its charge to a neutral, converting the neutral into an ion and the ion into a neutral. Despite the transfer of an electron, the heavy particles' velocities are not changed much. If the ion was moving fast and the neutral was "slow" before the collision, then after the collision the neutral will be fast and the ion will be slow (Fig. G.2).

Chemisorption.   Attachment of a molecule to a surface by formation of a chemical bond.

Child Law.   A model for a sheath that assumes that ions enter the sheath with little energy and fall through the sheath without collisions. The potential in the sheath varies as $x^{4/3}$ and the density varies as $x^{-2/3}$, where x is the distance from the plasma–sheath boundary.

Closure.   See Fluid Equations. A simplifying assumption used so that the fluid equations can be solved, without having to use an infinite number of fluid equations.

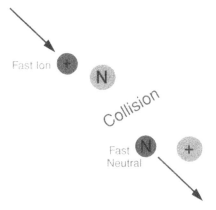

Fig. G.2. A charge-exchange collision of an fast ion with a neutral, producing a slow ion and a fast neutral.

Corona Discharge.   A type of discharge formed at around atmospheric-pressure, which is used for rapidly treating large areas of material.

Coulomb Collisions.   The predominantly long-range collisions between charged particles, due to their Coulomb repulsion or attraction of each other. The cumulative effect of the many long-range Coulomb collisions tends to produce more overall change in velocity than the fewer short-range Coulomb interactions, even though the deflection in each long-range Coulomb interaction is small. Coulomb interactions are largely cut off at distances beyond the Debye length.

Cumulative Distribution Function. $F_X(x)$ of a random variable X is the probability of that variable X being less than some value x.

CVD (Chemical Vapor Deposition).   The deposition of a film on a surface by chemical reactions in a gas and on the surface, where the reactants are all at (similiar or) the same temperature(s) and that temperature is high enough to drive the chemical reactions.

Dark Space.   One of the regions of a dc discharge where little or no light is emitted, including the Crookes dark space (or cathode fall), the Faraday dark space, and the anode dark space.

Debye Length.   The characteristic length scale λ_D on which electric fields decay in a plasma. The decay in electric field is due to shielding by the more mobile species, which are usually electrons. If there are more electrons the shielding is more effective, so λ_D is smaller. If the electrons become hotter they can escape the field better making the shielding less effective and λ_D larger.

Desorption.   Inverse process to adsorption, when a molecule detaches its bond with a surface.

Detailed Balance.   The relation between the rate constants for a reaction and for the reverse reaction. It implies that in thermodynamic equilibrium the rate of the forward reaction equals the rate of the reverse reaction.

Dielectric Relaxation Time.   See Plasma Frequency.

Differential Cross Section.   The derivative of the cross section with respect to solid angle, where the angle is the angle through which particles scatter. The contribution $d\sigma$ to the cross section σ, of collisions that scatter into the range of angles $d\theta$ near the angle θ, divided by $2\pi \sin\theta d\theta$.

Dipole Radiation.   When an atom has an electric dipole moment, it can emit dipole radiation, in a fashion that resembles an oscillating classical electric dipole.

Direct Simulation Monte Carlo.   Monte Carlo (MC) simulation where the collisions are not with a "background" gas, which does not respond to the collision, as is the case in most Monte Carlo simulations. Instead the gas being collided with is itself handled in the MC method and momentum is conserved [157].

Distribution Function.   See also Probability Density Function. The number of particles per unit of each of all the independent variables in phase space. The distribution function provides information about the number density in physical space as well as the fraction of the particles having particular speeds or velocities at each point in space.

Double Probe.   A pair of Langmuir probes, both of which have one end inserted in a plasma, with their other ends connected to each other through a voltage source.

Drift Velocity. The average motion of particles, on which random (usually thermal) motion is superimposed.

Druyvesteyn Distribution. Distribution function proportional to $e^{-\alpha v^4}$. Expected to occur for charged particles in a constant electric field when the cross section is also constant.

Dust. Macroscopic particles that form in a plasma and that may land on the surface being processed, potentially contaminating or damaging the surface.

E × B Drift. In the presence of both electric and magnetic fields, charged particles have a "drift velocity" $\mathbf{v_E} = \mathbf{E} \times \mathbf{B}/B^2$. Using $\mathbf{v_E}$ in the Lorentz force gives $\mathbf{F} = q(\mathbf{E} + \mathbf{v} \times \mathbf{B}) = q(\mathbf{E} + (\mathbf{E} \times \mathbf{B}) \times \mathbf{B}/B^2) = q(\mathbf{E} - \mathbf{E} + (\mathbf{E} \cdot \mathbf{B})\mathbf{B}/B^2) = q E_{\parallel}$, where $E_{\parallel}$ is the component of $\mathbf{E}$ parallel to $\mathbf{B}$; $E_{\parallel} = (\mathbf{E} \cdot \mathbf{B})\mathbf{B}/B^2$.

Einstein Relation. The relationship, $D = \frac{k_B T}{q}\mu$ between the mobility and the diffusion coefficient, which is derived from the expression for the flux per unit area of charged particles, $\Gamma = n\mu\mathbf{E} - D\nabla n$, which is zero in conditions of thermodynamic equilibrium, and from the Boltzmann expression for the equilibrium density, $n = n_o \exp(-\frac{q\Phi}{k_B T})$.

Elastic Collision. A collision in which no translational kinetic energy is converted into the internal energy of the colliding particles. The total translational kinetic energy, summed over all the particles, is the same before and after the elastic collision. When an electron collides elastically with a neutral, the much greater mass of the neutral means that the neutral is deflected very slightly but the electron can be deflected a great deal. Since the neutral is deflected very little, its kinetic energy also changes very little. Consequently, the electron's kinetic energy changes very little in elastic collisions with neutrals. Electron–neutral elastic collisions are exceptional in this regard; elastic collisions between particles of similar mass typically involve significant exchange of translational kinetic energy.

Electron Cyclotron Resonance. The resonance between electron gyromotion and an appropriate microwave electromagnetic field, which can be used to heat electrons.

Electropositive (negative) Plasma. An electropositive plasma is composed mainly of positive ions and electrons. An electronegative plasma contains similar numbers of positive and negative ions and has relatively few electrons.

Elliptic Equations. Elliptic equations are second-order linear partial differential equations such as Laplace's equation, having complex characteristics. Boundary conditions for an elliptic equation can be Dirichlet (which specify the dependent variable) or Neumann (which specify the derivative of the dependent variable in the direction normal to the boundary). In either case the boundary condition should be specified on a closed boundary, that is, one that runs all around the region being studied (although part of the closed boundary could be at infinity) [2].

Emissive Probe. A hot wire probe that emits electrons.

E/N. See Reduced Electric Field.

Energy Cascade. When turbulence is generated in a situation such as in a wind tunnel, the eddies upstream generate smaller ones further downstream. The energy is fed from low wave numbers to high wave numbers, until the eddies are small enough for viscosity to make dissipation effective.

Ensemble.   A large number of fictitious systems, with one system in every possible state of a real system. Averages done over the ensemble are expected to represent time averages of the real system's behavior.

Epitaxy.   Growth of a crystal on top of another crystal.

Ergodic State.   A state of a system, which is not periodic and which has a finite expected time between occasions when the system visits the state, is called an ergodic state.

Ergodic Theorem.   An ergodic theorem describes the conditions when a time average converges, when the average is done over a long time. Usually we will be interested in cases where the average converges to the ensemble average.

ESD.   Electrostatic Discharge. Usually refers to breakdown of a gate dielectric layer, due to charge build-up during processing.

Etching.   Removal of material from a surface, by chemical reaction or physical sputtering.

Expert System.   A computer program that acts as an interface to a data base containing extensive information on the topic on which the system is "expert."

Explicit Scheme.   Finite difference scheme where the "new" values of the variables can be written explicitly in terms of the old values. In a time evolution problem described using an explicit scheme the change in each variable is proportional to Δt. This is not accurate for large Δt. Explicit schemes may go unstable if too large a time step is used.

Factorial Design.   An experimental design where the independent variables are set to "low" and "high" values under different experimental conditions [5] (Fig. G.3).

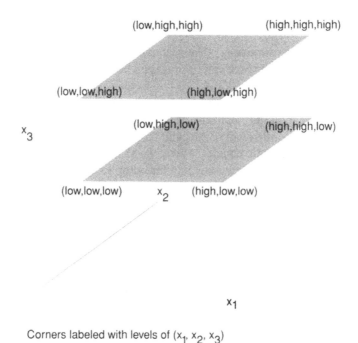

Corners labeled with levels of (x_1, x_2, x_3)

Fig. G.3. The layout of a factorial design that has three independent variables.

A full factorial design uses all possible permutations of low and high settings. A partial factorial attempts to provide the same information with a judicious choice of a subset of all the possible low and high settings.

Faraday Shield. An arrangement of conductors designed to shield radio frequency fields, of some or all polarizations, out of a region.

Field Line. For any vector field (which is a vector that exists throughout a region), a field line can be defined by starting at any point and moving a short distance in the direction of the vector at that point. At the new point that has been moved to, the direction of the vector is found and another short distance is moved in that direction. Repeating this process many times, we obtain the field line. Field lines for electric and magnetic fields are frequently used.

Finite Differences. The representation of derivatives by discrete approximations, when the dependent variable is known at locations separated by finite small differences in the independent variable(s). The difference in the dependent variable at successive locations, divided by the distance between locations, would be an example of a finite difference approximation to a first derivative. In other words, $\frac{dV}{dx}$ can be approximated as $\frac{V_{i+1}-V_i}{\Delta x}$, where V_i is the value of V at the ith point at which V is known and Δx is the difference in x between points i and $i+1$.

Floating Potential. Potential at which a probe in a plasma draws no net current.

Fluid Equations. Usually refers to the equations obtained by treating the plasma as one or more fluids with average properties at each point in space, such as mass density, average velocity, temperature, etc. Each of these quantities is governed by a fluid equation. The set of fluid equations is not "closed" – the lower order equations cannot be solved without information contained in higher order equations. The fluid equations are in principle exact. The choice of closure (which is an assumption introduced instead of solving the higher order equations, such as using an Ohm's law to find the current density) introduces an approximation that is not possible to justify in general.

Flux. The number of particles (or the amount of some other quantity) crossing an area per second. The flux per unit area Γ is the number of particles crossing a unit area per second.

Flux Coordinates. A coordinate system based on the magnetic field lines [57]. Distance along a field line is usually one of the coordinates.

Fowler–Nordheim Tunneling. Mechanism for electron emission from a surface, which depends on quantum-mechanical tunneling.

Free Radical. See Radical.

Gamma Regime. Regime of operation of a discharge where ionization is primarily caused by secondary electrons emitted from the electrodes.

Gaussian Distribution, or Normal Distribution. Distribution having the probability density function $f_X(x) = (\sqrt{2\pi}\sigma)^{-1}\exp(-(x-\mu)^2/2\sigma^2)$, where μ is the mean and σ^2 the variance.

Guiding Center. The center of the gyration of a charged particle in a magnetic field.

Gyromotion. The motion of a charged particle as it orbits around a magnetic field line. 2π times the frequency of the orbit is called the "gyrofrequency," $\omega_c = \frac{eB}{m}$ and the radius of the orbit $r_g = v_\perp/\omega_c$ is called the gyroradius, or Larmor radius.

H-Theorem. States that the collision operator in a kinetic equation must lead to increasing entropy, or constant entropy in equilibrium.

Hydrodynamics. See Fluid Equations. Literally means the movement of water. Hydrodynamic description of a plasma means treating the plasma as behaving like a fluid such as water, and using equations to describe the plasma that resemble the equations describing such a fluid.

Hyperbolic Equation. A hyperbolic equation is a second-order linear partial differential equation that resembles a wave equation, $\frac{\partial^2 V}{\partial x^2} - \frac{1}{u^2}\frac{\partial^2 V}{\partial x^2} = 0$. It has real "characteristics." In the case of the wave equation, characteristics correspond to wave fronts. If $\lambda_1 = x - ut$ and $\lambda_2 = x + ut$, then $\lambda_1 =$ constant is a wave front for a wave traveling in the x direction and $\lambda_2 =$ constant is a wave front for a wave traveling in the negative x direction. The solution of the wave equation depends on λ_1 and λ_2 through $V = f_1(\lambda_1) + f_2(\lambda_2)$. f_1 and f_2 are arbitrary functions of λ_1 and λ_2. Boundaries used to specify boundary conditions must cross all values of λ_1 and λ_2, and they must cross them only once. The boundary condition would be specified at a single value of λ_1 or λ_2 if the boundary used coincided with a wave front. There would then be no information available about conditions at the other wave fronts. The wave equation needs "Cauchy" boundary conditions; the value of V and the slope of V normal to the boundary must be specified. In other words, Cauchy boundary conditions involve specifying V and $\frac{\partial V}{\partial t}$ when the normal to the boundary is in the "t direction" and the normal derivative is $\frac{\partial V}{\partial t}$.

Impact Parameter, b. Suppose a moving particle is approaching a stationary particle. b is the perpendicular distance between the stationary particle and the initial trajectory of the moving particle.

Implicit Scheme. Finite difference scheme where some of the derivatives (usually derivatives with respect to time) are approximated relatively accurately. The "new" values of the variables at a point in space are not given explicitly only in terms of the old values (at that and adjacent points in space) but are given in terms of old values and new values at that and adjacent physical locations. An implicit scheme often requires solution of a matrix equation to find the new values. It is usually more stable than an explicit scheme.

Inductive Discharge. A discharge where the power is supplied primarily by passing current through a primary winding close to the plasma and causing a current to flow through the plasma, which acts as the secondary winding.

Inelastic Collision. A collision in which translational kinetic energy is converted into internal energy of the colliding particles.

Integrating Factor Method. Solution of an equation of the form $\frac{dx}{dt} + \alpha x = f(t)$ by writing the left-hand side as $\exp(-\alpha t)\frac{d}{dt}(\exp(\alpha t)x)$.

Ion Pumping. The collisional transfer to neutrals of momentum from ions that are moving out of the plasma. The neutrals consequently also tend to be pushed out of the center of the plasma.

Isotropic. The same in all directions. Isotropic etching refers to etching that happens at the same rate in all directions. An isotropic distribution has the same number of particles traveling in all directions, at any given point in space. Anisotropic means not the same in all directions.

Kinematic Viscosity. $v = \mu/\rho$ is the ratio of the viscosity to the density.

Kinetic Equation. Equation for the time evolution of the distribution function f, of the form $\frac{df}{dt} = C(f)$. The time derivative is evaluated along a particle trajectory in the phase space, and $C(f)$ is the collision operator.

Krook Collision Operator. Describes a collision process where particles are removed from the distribution f at a rate given by the collision frequency v and replaced at zero velocity. The removal is described by a collision term equal to $-vf$.

Landau Damping. Collision less damping of electrostatic waves, which causes the distribution function to oscillate as a function of the particle speed.

Langevin Cross Section. Cross section for collision between an ion and a polar (i.e., polarizable) molecule.

Langmuir Probe. A wire inserted into a plasma and biased at different voltages. The variation with voltage of the current drawn is used to indicate plasma parameters.

Larmor Radius. See Gyromotion.

Le Chatelier's Principle. States that when an external constraint is placed on a system in equilibrium, the equilibrium will shift in such a direction as to tend to remove the constraint. If a liquid and its vapor are in contact and the pressure is increased by decreasing the volume, the system decreases the pressure again by condensing some of the vapor. Similarly, in a chemical equilibrium between gaseous reactants and products, if more of one of the products is added, the equilibrium will shift to use up some of the gas that was added and produce more of the reactants.

Lithography. The process of optical pattern-transfer onto a material, using photoresist and subsequent etching of the material.

Local Heating. Refers to a situation where electrons do not move quickly between regions of the discharge (in some sense) having different electric fields. Instead, they reach equilibrium with the local electric field so their mean energy and other behaviors depend on the local E/N.

Local Thermodynamic Equilibrium, LTE. The equilibrium achieved in some plasmas, usually thermal plasmas, between locally occurring processes. Radiation processes are usually the primary cause of lack of overall equilibrium when a system is in LTE.

Loss Cone. A range of velocities where particles are lost rapidly. Suppose particles of the same kinetic energy and hence the same speed v approach an electric field that repels them (as happens to electrons near a sheath). Only those particles with a high enough velocity $v_\perp$ in the direction toward the higher potential can get over the potential. $v_{\perp min}$ is the minimum $v_\perp$ to get over the potential. $v_\perp$ is greater than or equal to $v_{\perp min}$ in a range of angles $\Delta\theta$ on either side of the direction with the steepest potential gradient. This range of angles is the loss cone and is given by $\cos \Delta\theta = v_{\perp min}/v$. (Similarly, when a magnetic field is present and its strength varies along the magnetic field lines, some particles cannot get past the maximum in the field. The magnetic moment μ of particles gyrating about the fieldlines is approximately conserved between collisions, in some circumstances. Since $\mu \equiv mv_\perp^2/2B$, where $v_\perp$ is now the velocity perpendicular to the field line, then $\frac{1}{2}mv_\perp^2 = \mu B$. When B increases so does $v_\perp$, but the change

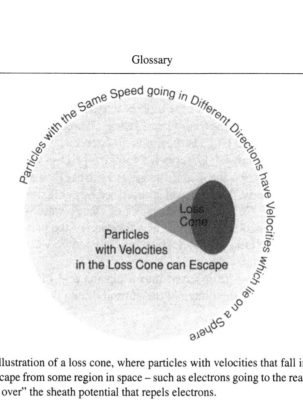

Fig. G.4. Illustration of a loss cone, where particles with velocities that fall in the loss
cone can escape from some region in space – such as electrons going to the reactor wall,
by "getting over" the sheath potential that repels electrons.

in B does not lead to a change in the kinetic energy so $v_\parallel$ decreases. If $v_\parallel$ goes to
zero the particle bounces. At a given speed v the range of values of $v_\parallel$ that allow
particles to escape is also a loss cone (Fig. G.4).)

Magnetic Mirror. A region where magnetic field lines crowd together so that
particles moving along field lines into the "mirror" experience an increase in
magnetic field strength. Conservation of the "magnetic moment" $\mu = m v_\perp^2 / 2B$
means that $\mu B = \frac{1}{2} m v_\perp^2$ gets bigger when B gets bigger. The total kinetic energy
$T = \frac{1}{2} m (v_\perp^2 + v_\parallel^2)$, so if the potential energy is constant and $\frac{1}{2} m v_\perp^2$ increases then
$\frac{1}{2} m v_\parallel^2$ must decrease. If $v_\parallel$ reaches zero it reverses sign and the particle bounces
in the mirror.

Magnetic Moment. See Adiabatic Invariant and Magnetic Mirror. The magnetic
moment μ of a current loop is the current multiplied by the area of the loop.
For a charged particle gyrating around a field line, the current is the charge q
multiplied by the frequency of gyration $f_c = \frac{\omega_c}{2\pi}$ and $\omega_c = \frac{qB}{m}$. The area of the
loop is $\pi r_\perp^2$, where $r_\perp$ is the gyroradius; $r_\perp = v_\perp / \omega_c$, so

$$\mu = q f_c \pi r_\perp^2 = \frac{q \omega_c}{2\pi} \pi \frac{v_\perp^2}{\omega_c^2} = \frac{q}{2} \frac{v_\perp^2}{\omega_c} = \frac{m v_\perp^2}{2B}.$$

Magnetic Probe. Measures the rate of change of magnetic flux through a small
loop of wire.

Magnetohydrodynamics, MHD. See Hydrodynamics. MHD refers to the behavior of a conducting fluid in a magnetic field. Describing a plasma using MHD
can sometimes imply that it is behaving like a fluid such as liquid mercury would,
depending on the choice of closure used to solve the equations.

Magnetron. A dc discharge that uses a magnetic field to improve the confinement of electrons. Magnetrons are frequently used to generate plasmas for use in sputter deposition of films [6, 7].

Mask. A layer of material formed on a surface, usually to protect parts of the surface from being etched.

Matching Network. A lossless network placed between the power supply and the discharge to ensure maximum power transfer.

Matrix Sheath. Sheath with an ion density that is constant in space, so the electric field varies linearly with position.

MBE. Molecular Beam Epitaxy. Technique used to deposit high-quality epitaxial films by vacuum evaporation (sublimation) of material in crucibles.

MEMS. MicroElectroMechanical Systems. "Microscopic" machines; their fabrication and use. The machines are made using the techniques of microelectronics.

Metastable Atom. An atom in an excited state that cannot decay by a rapid transition such as by an electric dipole transition.

Method of Averaging. Calculation of the long-term variation of a variable, by averaging the nonlinear effects of rapid oscillations in the variable.

Method of Characteristics. A method for solution of equations with real characteristics; see Hyperbolic Equations. For a collisionless kinetic equation, an initial condition can be advanced in time using the fact that the distribution function is constant along the trajectory of a particle in phase space.

MOCVD. Metal Organic Chemical Vapor Deposition. Technique used for preparation of high-purity, epitaxial layers of semiconductors, especially III–V materials.

Moment of the Distribution Function. The integral, over velocity, of the distribution function multiplied by the velocity raised to an integer power: $\int f v^n d\mathbf{v}$.

Momentum Transfer Cross Section. The total cross section is $\sigma = \int \frac{d\sigma}{d\Omega} 2\pi \times \sin\theta d\theta$. $\frac{d\sigma}{d\Omega}$ is the differential cross section. To allow for the fact that a fraction of a particle's momentum, $\alpha = (1 - \cos\theta)$, is lost in any collision, the momentum transfer cross section is defined with an extra factor of α in the integrand, compared to the expression above.

Monte Carlo Method. Simulation of a system using random numbers to determine the evolution of the system. In plasma applications the system usually comprises a large number of particles. By following individual particles and generating random numbers to determine when the particles have collisions, the behavior of a gas can be approximated. Monte Carlo (MC) methods are often relatively straightforward to set up, but the results must be interpreted carefully, in part because the number of particles in any region of phase space is subject to statistical fluctuations. If the number of particles in a subregion of the simulation is N_p then the expected uncertainty in N_p is of the order of $\sqrt{N_p}$ (in the absence of a self-consistent electric field) so the fractional error is of the order $\sqrt{N_p}/N_p = N_p^{-1/2}$. This can lead to a need for a very large number of simulation particles and can make MC methods computationally expensive.

Negative Feedback. The tendency of a system to respond to an applied stimulus by (partially) removing the stimulus. See Le Chatelier's Principle.

Negative Glow. The region of a dc discharge after the cathode fall.

Neural Net. A computer program that "learns" from experience, in a fashion intended to resemble human learning in some respects. Neural nets have been used in plasma processing to learn how etching is correlated to the externally controllable settings of the reactor.

Nonlocal Heating. Refers to a situation where electrons in a discharge can move around the discharge rapidly enough so that their mean energy does not depend only on the local E/N. Instead they can, for instance, be heated in a strong field region and move rapidly into a region with a weaker field where they may cause ionization, etc.

Numerical Diffusion. When a density is described using a finite difference scheme, the mesh spacing is nonzero and so the particle positions are only known to an accuracy equal to the mesh spacing. At each time step particles are moved to reflect their motion in a time step and put back on the mesh. Their positions where they are replaced are inaccurate (even assuming they were right at the start of the time step) by an amount that can be as large as the mesh spacing. This inaccuracy causes an artificial spreading of the density, known as "numerical diffusion."

Nyquist Criterion. Several different Nyquist criteria exist. In signal processing, the Nyquist criterion states that it is necessary to sample an oscillation at least twice per period to resolve it. If the sampling happens every τ_s seconds then the shortest-period oscillation resolved has period $2\tau_s$. The highest frequency resolved is thus $(2\tau_s)^{-1}$.

Ohmic Heating. Heating that occurs due to particles having a drift velocity $\mathbf{v_d}$ caused by the local electric field $\mathbf{E}$.

Overetch. Etching past what is the desired end point of the etching over most of the wafer, to ensure end point is reached at all locations on the wafer.

Parabolic Equations. Parabolic equations are second-order linear partial differential equations that resemble the diffusion equation. They have only one set of characteristics because in a diffusion process the equation can be solved if we know the initial condition but (in general) not if we only know a final condition. The solutions decay to a steady state that does not usually depend on the initial transient, unlike a wave equation – see Hyperbolic Equations. Dirichlet boundary conditions should be used (that is, the dependent variable specified) at the boundary (in time) that corresponds to the earliest time considered.

Particle Simulation, or Particle-In-Cell Method. Refers to computer simulations of plasmas (or other gases) that represent the plasma as large numbers of simulation particles [55]. The equations of motion of each simulation particle are integrated in time. After each time step the charge density ρ (and if necessary the current density $\mathbf{J}$) are found from the behavior of the simulation particles, treating each simulation particle as representing a large number N_r of real particles (where typically $N_r \sim 10^{12}$). The new electromagnetic fields are found from ρ and $\mathbf{J}$ so that the next time step of the particle motion can be performed, and so on.

PECVD/PACVD. Plasma Enhanced (Assisted) Chemical Vapor Deposition. A process where plasma assists in the deposition of a film from gaseous precursors. Usually the plasma electrons provide the energy to activate the chemical reactions, which produce the film.

Penning Ionization.   Ionization that occurs when two metastable atoms collide and one is ionized while the other returns to the ground state.

Phase Space.   A set of independent variables used in writing a kinetic equation and in studying particle motion, including at least one spatial coordinate and at least one velocity or energy.

Photoresist.   A material that is deposited onto a surface in thin film form and that can be photographically "developed." The effect of light on the photoresist may be to strengthen or weaken the photoresist with respect to etching. In either case the light is used to form a pattern on the resist. The resist then acts as a mask for etching the material underneath the resist.

Physisorption.   Attachment of a molecule to a surface by polarization (van der Waals) force.

Pitchfork Bifurcation.   When an elastic strut (a rod) is under a compressive force F, it may stay straight or it may bow sideways. If the maximum sideways displacement d is plotted against F it looks like a pitchfork. Below a critical force F_c the only solution is $d = 0$. For $F > F_c$, there are a pair of stable solutions, $d = \pm d_c(F)$, while $d = 0$ is an unstable solution. The displacement obeys an equation like $d^3 - \alpha(F - F_c)d = 0$. Another example of a pitchfork bifurcation occurs when a fluid flow has a sudden expansion. As the Reynolds number increases the flow goes from being symmetric to one with an asymmetric recirculation pattern.

Plasma.   Usually refers to any collection of charged particles that exhibit collective behavior in which their long-range interactions through the electromagnetic field are important.

Plasma Approximation.   The number of charged particles in the Debye sphere is assumed to be very large; $n\lambda_D^3 \gg 1$.

Plasma Frequency and Plasma Period.   If the electrons in a plasma are all moved in the same direction relative to the ions, an electric field is set up that pulls them back. If there are no collisions the electrons pick up kinetic energy as they are pulled back, and they overshoot the position where the force is zero and oscillate on either side of it. The frequency of the oscillation is the plasma frequency (or electron plasma frequency)

$$f_{\mathrm{p}} = \frac{1}{2\pi}\left(\frac{n_e e^2}{m\varepsilon_o}\right)^{1/2}.$$

The plasma period $T_{\mathrm{p}} = 1/f_{\mathrm{p}}$. If collisions quickly remove the kinetic energy that the electrons pick up, they do not overshoot but return to zero displacement. Their displacement decays exponentially on the time scale of the dielectric relaxation time

$$t_{\mathrm{d}} = \frac{\varepsilon_o}{N\mu q}.$$

Plasma Gun.   A plasma source, an early version of which was made from a stack of titanium washers. The titanium will react with hydrogen to form titanium hydride at about 1,000°C. Electrons are accelerated from a source through about

1 kV and strike the washers, heating the surface and releasing hydrogen, which becomes ionized.

Plasma Parameter, g.   The number of charged particles in a Debye sphere. A Debye sphere is a sphere with radius of the Debye length λ_D; so $g = n\frac{4}{3}\pi\lambda_D^3$, where n is the charged particle density. Since λ_D varies as $(T_e/n_e)^{1/2}$ in most plasmas, where electrons are the species responsible for Debye shielding, the plasma parameter varies as $g \sim T_e^{3/2}/n_e^{1/2}$. The plasma parameter is large in a system of charged particles where collective effects are important (i.e., in a plasma).

Plasma Polymerization.   Polymerization that takes place because of the presence of plasma. The polymer formed usually exhibits a relatively high degree of cross-linking so it is not a true polymer since a true polymer consists of a chain of identical units. Whether the polymer forms in the gas phase or on the surface is not always clear at present.

Plasma Potential.   Potential of the plasma in steady state relative to a wall with which it is in contact.

Polarization Scattering.   A charged particle can cause an atom to become polarized, if there is enough time for the atom to respond as the charged particle passes by. Once the atom is polarized it attracts the charged particle, so they can scatter off each other. This short-range scattering process is the main cause of ion–neutral collisions and of collisions of low-energy electrons with neutrals.

Ponderomotive Force.   A radio frequency, inhomogeneous electric field E can exert a time-averaged force on a charged particle, proportional to $-\nabla E^2$, called the ponderomotive force.

Positive Column.   A region of a dc discharge typically encountered in long cylindrical discharges that is overall positively charged and that is nearly homogeneous in the z direction. Particle losses in the radial direction balance ionization caused by the electric field in the z direction.

Presheath.   A quasineutral region next to the sheath where electric fields accelerate ions before they enter the sheath.

Probability Density Function (pdf). $f_X(x)$ is the density of probability in the sense that the probability of x being between a and b is the integral of $f_X(x)$ from a to b, $\int_a^b f_X(x)dx$. This implies that the cumulative distribution function $F_X(x) = \int_{-\infty}^x f_X(y)dy$, and $f_X(x) = \frac{dF_X(x)}{dx}$.

Probe.   A small conductor inserted into a plasma and connected to a circuit that measures electrical characteristics of the plasma. The current drawn by the probe at different voltages applied to the probe is used to characterize the plasma. Probes are usually simple to use but their measurements are often difficult to interpret since they can alter the state of the plasma.

PSII or PIII.   Plasma Source (or Immersion) Ion Implantation [14, 15, 16]. A method for implanting ions relatively uniformly into the surface of an object, by inserting the object into a plasma and then biasing the object to a large negative voltage. The sheath expands as the voltage is ramped down until ions from a large fraction of the plasma are collected and hit the target with an energy corresponding to the sheath potential. PSII has been applied to implanting dopants into semiconductors to form very shallow junctions.

Q-Machine.   A device designed to contain plasma produced by ionization of alkali metal atoms that touch a hot tungsten surface. The plasma was at first thought to be Quiescent.

Quasineutrality.   Most of a plasma is usually quasineutral, meaning the total positive charge per unit volume is approximately equal to the total negative charge per unit volume. The quasineutral approximation when it applies can be used to estimate either n_e or n_i from the other, but it is not accurate enough to use in Poisson's equation.

Radical.   A molecule with an unpaired outer electron, so that the electron is available to form a covalent bond. Radicals are consequently highly chemically reactive.

Random Process.   A number that evolves randomly with time or with some other independent variable can be thought of as a "family" of random variables, each labeled by the value of the time to which it applies. Each possible time evolution of the number can also be thought of as a particular realization of the random process. The random process itself consists of the family of random variables indexed by the time.

Random Variable.   A random variable "assigns" a number to the outcome of a random experiment. The outcome of tossing a coin some number of times (which is a random experiment) can be assigned a number by counting the number of heads; hence the number of heads is a random variable. The randomness in the variable is caused by the randomness in the experiment.

Rayleigh–Taylor Instability.   The instability expected when a pair of adjacent fluids can reduce their potential energy by changing places – such as when a dense liquid is in a layer above a less dense liquid.

Reduced Electric Field, or E/N.   In cases where electrons are primarily heated locally, their distribution may be characterized by the amount of energy $\Delta \varepsilon$ they acquire from the electric field E in traveling a mean free path, $\lambda = 1/N\sigma$. Since $\Delta \varepsilon$ is proportional to $E\lambda$ and hence to E/N the properties of the distribution are often plotted as functions of the reduced electric field E/N.

Resonant Charge Exchange.   A charge-exchange collision between an ion and its own neutral, wherein the same species are present before and after the collision. Resonant charge exchange usually has a relatively large cross section.

Reynold's Number, R.   Roughly the ratio of inertial forces to viscous forces.

Saha Equation.   "Equation of State" for a plasma in local thermodynamic equilibrium. It gives the degree of ionization, in terms of the temperature.

Selectivity.   The ability of an etching process to etch one material preferentially over another material.

Sheath.   A region of strong electric fields, usually set up by electrons that charge a boundary (which is in contact with a plasma) negatively. The sheath electric field close to the negative boundary repels other electrons from the boundary and attracts positive ions, so the sheath is usually a positively charged region.

Small-Angle Scattering.   Coulomb collisions, between charged particles, deflect particles mostly because of the cumulative effect of many small-angle scatters, (each of which by itself only deflects the particles through a small angle).

Sputtering.   The process in which energetic ions or other particles knock atoms out of a solid material and into the gas phase.

Stiff Equations.   A system of equations where a set of spatial or time scales, which have very large differences between them, must be resolved. For example, plasma models tend to be stiff because the electron time scale is much shorter than the ion time scale, and processes happening on each of the time scales must be described properly when the model equations are solved.

Stochastic Heating.   In some plasmas, energetic electrons enter and leave a region of strong oscillatory electric fields multiple times, having a random initial phase relative to the phase of the field each time they enter the region. The electrons' energy may tend to increase on average due to the randomness of the initial phase. This process is referred to as stochastic heating.

Stochastic Process.   Same as Random Process.

Swarm Experiment.   Experiments on the behavior of electrons in spatially uniform electric fields are called "swarm" experiments. The characteristics of the electrons, such as mean energy, drift velocity, ionization rate, and so forth, typically are tabulated for various neutral gases as functions of E/N, where E is the electric field strength and N is the neutral particle number density.

TEOS. $Si(OC_2H_5)_4$. A liquid, the vapor of which is used to grow SiO_2 films.

Thermal Plasma.   A type of plasma formed at relatively high neutral pressures. In thermal plasmas the neutral species are usually at a high temperature and the charged and neutral species can be described meaningfully as fluids at well-defined temperatures. Thermal plasmas are used for spraying films and in plasma torches, and they occur in sparks formed in circuit breakers.

Thermionic Emission.   Emission of electrons from a surface caused by heating the surface.

Townsend Coefficient.   The flux of electrons falling in a constant electric field increases as $e^{\alpha z}$, where α is the first Townsend coefficient.

Turbulence.   A state of continuous instability.

Two-Stream Instability.   An electrostatic instability that occurs when a plasma has two (or more) components with a nonzero velocity relative to each other.

Uniform Random Variable.   The probability density function (pdf) of a uniform random variable is $f_X(x) = \frac{1}{b-a}$, for x between some a and b, and the pdf is zero elsewhere.

Variance.   The variance of a random variable X is the expected value of $(X - \mu)^2$, where μ is the mean (or expected) value of X.

Vlasov Equation.   The kinetic equation in the absence of collisions, and with the force on the particles being the Lorentz force due to the electromagnetic fields.

Bibliography

[1] Sun Tzu. *The Art of War*. Oxford University Press, Oxford, 1963.

[2] P. M. Morse and H. Feshbach. *Methods of Theoretical Physics*. McGraw-Hill, New York, 1953.

[3] S. Chapman and T. G. Cowling. *The Mathematical Theory of Non-Uniform Gases*. Cambridge University Press, Cambridge, 1952.

[4] D. C. Montgomery. *Design and Analysis of Experiments*. Wiley, New York, 1997.

[5] G. E. P. Box, W. G. Hunter, and J. S. Hunter. *Statistics for Experimenters, An Introduction to Design, Data Analysis, and Model Building*. Wiley Series in Probability and Mathematical Statistics. Wiley, New York, 1978.

[6] D. M. Manos and D. L. Flamm. *Plasma Etching: An Introduction*. Academic, San Diego, 1989.

[7] S. M. Rossnagel, J. J. Cuomo, and W. D. Westwood. *Handbook of Plasma Coating Technology*. Noyes Publications, NJ, 1990.

[8] J. A. Thornton. *Journal of Vacuum Science and Technology*, 12:830, 1975.

[9] B. A. Movchan and V. A. Demchishin. *Fizika Metall*, 28:653, 1969.

[10] J. B. Adams. *Nucleation and Growth of Thin Films*. PhD thesis, University of Wisconsin at Madison, 1987.

[11] T. J. Lenosky, J. D. Kress, I. Kwon, A. F. Voter, B. Edwards, D. F. Richards, S. Yang, and J. B. Adams. *Physical Review B*, 55:1528, 1997.

[12] S. H. Yang, D. A. Drabold, and J. B. Adams. *Physical Review B*, 48:5261, 1993.

[13] C. L. Liu, J. M. Cohen, J. B. Adams, and A. F. Voter. *Surface Science*, 253:334, 1991.

[14] J. R. Conrad, J. L. Radke, R. A. Dodd, F. J. Worzala, and N. C. Tran. *Journal of Applied Physics*, 62:4591, 1987.

[15] M. Shamim, J. T. Scheuer, and J. R. Conrad. *Journal of Applied Physics*, 69:2904, 1991.

[16] J. Tendys, I. J. Donnely, M. J. Kenny, and J. T. A. Pollock. *Applied Physics Letters*, 53:2143, 1988.

[17] G. J. Parker, W. N. G. Hitchon, and E. R. Keiter. *Physical Review E*, 54:938, 1996.

[18] J. P. Biersack and L. G. Haggmark. *Nuclear Instruments and Methods*, 174:257, 1980.

[19] J. Albers. *IEEE Transactions on Electron Devices*, 32:1930, 1985.

[20] W. Bohmay, A. Burenkov, J. Lorenz, H. Ryssel, and S. Selberherr. *IEEE Transactions on Semiconductor Manufacturing*, 8:402, 1995.

[21] L. A. Christel, J. F. Gibbons, and S. Mylroie. *Journal of Applied Physics*, 51:6176, 1980.

[22] M. D. Giles and J. F. Gibbons. *IEEE Transactions on Electron Devices*, 32:1918, 1985.

[23] J. Lindhard, V. Nielsen, M. Scharff, and K. Dan Vidensk. *Selsk. Mat. Fys. Medd.*, 36(10), 1968.

[24] O. B. Firsov. *Zh. Eksp. Teor. Fiz.*, 32:1464, 1957.

[25] O. B. Firsov. *Zh. Eksp. Teor. Fiz.*, 34:447, 1958.

[26] O. B. Firsov. *Zh. Eksp. Teor. Fiz.*, 36:1517, 1959.

[27] F. H. Eisen. *Canadian Journal of Physics*, 46:561, 1968.

[28] K. H. R. Kirmse, A. E. Wendt, S. B. Disch, J. Z. Wu, I. C. Abraham, J. A. Meyer, R. A. Breun, and R. C. Woods. *Journal of Vacuum Science and Technology*, B 14:710, 1996.

[29] K. M. Kramer and W. N. G. Hitchon. *Semiconductor Devices: A Simulation Approach*. Prentice-Hall, Upper Saddle River, NJ, 1997.

[30] Y. P. Raizer. *Gas Discharge Physics*. Springer-Verlag, Berlin, 1997.

[31] Y. P. Raizer and M. N. Shneider. *Radio Frequency Capacitive Discharges*. CRC, Roca Baton, FL, 1995.

[32] J. H. Ingold. *Gaseous Electronics*. Academic Press, New York, 1978.

[33] M. A. Lieberman and A. J. Lichtenberg. *Principles of Plasma Discharges and Materials Processing*. Wiley, New York, 1994.

[34] R. O. Dendy, ed. *Plasma Physics: An Introductory Course*. Cambridge University Press, Cambridge, 1993.

[35] R. O. Dendy. *Plasma Dynamics*. Oxford University Press, Oxford, 1990.

[36] R. A. Cairns. *Plasma Physics*. Blackie, Glasgow, 1985.

[37] B. N. Chapman. *Glow Discharge Processes: Sputtering and Plasma Etching*. Wiley, New York, 1980.

[38] S. C. Brown. *Basic Data of Plasma Physics*. Wiley, New York, 1959.

[39] C. L. Longmire. *Elementary Plasma Physics*. Wiley, New York, 1963.

[40] P. C. Clemmow and J. P. Dougherty. *Electrodynamics of Particles and Plasmas*. Addison-Wesley, London, 1990.

[41] T. G. Northrop. *The Adiabatic Motion of Charged Particles*. Interscience, New York, 1963.

[42] T. Tajima. *Computational Plasma Physics*. Addison-Wesley, 1989.

[43] N. A. Krall and A. W. Trivelpiece. *Principles of Plasma Physics*. McGraw-Hill, London, 1973.

[44] R. N. Franklin. *Plasma Phenomena in Gas Discharges*. Clarendon, Oxford, 1976.

[45] A. von Engel. *Ionized Gases*. Oxford University Press, Oxford, 1965.

[46] D. S. Rickerby and A. Matthews. *Surface Engineering: Processes, Characterization and Applications*. Blackie, Glasgow, 1991.

[47] S. P. Howlett, S. P. Timothy, and D. A. J. Vaughan. *Industrial Plasmas: Focussing UK Skills on Global Opportunities*. CEST, London, 1992.

[48] L. D. Landau and L. M. Lifshitz. *Fluid Mechanics. Course of Theoretical Physics*, vol. 6. Pergamon, Oxford, 1959.

[49] A. S. Monin and A. M. Yaglom. *Statistical Fluid Mechanics*. MIT Press, Boston, 1965.

[50] C. Y. Chow. *An Introduction to Computational Fluid Mechanics*. Wiley, New York, 1978.

[51] D. J. Triton. *Physical Fluid Dynamics*. Van Nostrand Reinhold, 1977.

[52] T. Mullin. *The Nature of Chaos*. Oxford University Press, Oxford, 1993.

[53] N. N. Bogolyubov and Y. A. Mitropolskii. *Asymptotic Methods in the Theory of Nonlinear Oscillations*. Gordon and Breach, New York, 1961.

[54] D. Potter. *Computational Physics*. Wiley, New York, 1973.

[55] R. W. Hockney and J. W. Eastwood. *Computer Simulation Using Particles*. McGraw-Hill, New York, 1981.

[56] W. Feller. *An Introduction to Probability Theory and Its Applications*. Wiley, New York, 1968.

[57] W. D. D'Haeseleer, W. N. G. Hitchon, J. D. Callen, and J. L. Shohet. *Flux Coordinates and Magnetic Field Structure*. Springer-Verlag, Berlin, 1991.

[58] M. M. Turner. *Physical Review Letters*, 71:1844, 1993.

[59] M. M. Turner. *Plasma Sources, Science and Technology*, 5:159, 1996.

[60] V. A. Godyak, R. B. Piejak, and B. M. Alexandrovich. *Plasma Sources, Science and Technology*, 3:169, 1994.

[61] Y. P. Song, D. Field, D. F. Klemperer. *Journal of Physics D: Applied Physics*, 23:673, 1977.

[62] V. A. Godyak, R. B. Piejak, and B. M. Alexandrovich. *IEEE Transactions on Plasma Science*, 19:660, 1991.

[63] K. Suzuki, S. Okudaira, N. Sakudo, and I. Kanomata. *Japanese J. Appl. Phys.*, 16:1979, 1977.

[64] J. S. Ogle. U.S. Patent 4 948 458, Aug. 14, 1990.

[65] D. K. Coultas and J. H. Keller. European Patent, publ. 0 379 828 A2, Jan. 8, 1990.

[66] H. N. Kucukarpaci, H. T. Saelee, and J. Lucas. *Journal of Physics D: Applied Physics*, 14:9, 1981.

[67] T. J. Sommerer, W. N. G. Hitchon, and J. E. Lawler. *Physical Review A*, 39:6356, 1989.

[68] J. Dutton. *J. Phys. Chem. Ref. Data*, 4:577, 1975.

[69] A. E. Wendt and W. N. G. Hitchon. *Journal of Applied Physics*, 71:4718, 1992.

[70] G. J. Parker, W. N. G. Hitchon, and J. E. Lawler. *Physics of Fluids B*, 5:646, 1993.

[71] V. I. Kolobov and D. J. Economou. *Plasma Sources, Science and Technology*, 6:1, 1997.

[72] V. I. Kolobov, D. P. Lymberopoulos, and D. J. Economou. *Physical Review E*, 55:3408, 1997.

[73] V. I. Kolobov and L. D. Tsendin. *Plasma Sources, Science and Technology*, 4:551, 1995.

[74] I. D. Kaganovich, V. I. Kolobov, and L. D. Tsendin. *Applied Physics Letters*, 69:3818, 1996.

[75] D. F. Beale. *Non-Local Electron Kinetics in an Inductively Coupled Plasma: 2D Model and Experimental Validation.* PhD thesis, University of Wisconsin at Madison, 1995.

[76] J. Hopwood. *Applied Physics Letters*, 62:940, 1993.

[77] S. M. Rossnagel and J. Hopwood. *Applied Physics Letters*, 63:3285, 1993.

[78] J. Asmussen, T. A. Grotjohn, P. Mak, and M. A. Perrin. *IEEE Transactions on Plasma Science*, 25:1196, 1997.

[79] J. E. Lawler, E. A. den Hartog, and W. N. G. Hitchon. *Physical Review A*, 43:4427, 1991.

[80] V. A. Godyak, R. B. Piejak, and B. M. Alexandrovich. *Journal of Applied Physics*, 73:3657, 1993.

[81] V. I. Kolobov and V. A. Godyak. *IEEE Transactions on Plasma Science*, 23:503, 1995.

[82] W. N. G. Hitchon, G. J. Parker, and J. E. Lawler. *IEEE Transactions on Plasma Science*, 22:267, 1994.

[83] J. R. Hollahan and A. T. Bell. *Techniques and Applications of Plasma Chemistry.* Wiley, New York, 1974.

[84] H. V. Boenig, ed. *Advances in Low Temperature Plasma Chemistry, Technology, Applications.* Technomics Publishing Co., Lancaster, PA, 1991.

[85] M. Shen. *Plasma Chemistry of Polymers.* Marcel Dekker, New York, 1976.

[86] H. V. Boenig. *Plasma Science and Technology.* Cornell University Press, Ithaca, NY, 1982.

[87] H. Yasuda. *Plasma Polymerization.* Academic Press, New York, 1985.

[88] R. d'Agostino. *Plasma Deposition, Treatment and Etching of Polymers.* Academic Press, New York, 1990.

[89] H. Biederman and Y. Asoko. *Plasma Polymerization Processes, Plasma Technology*, vol. 3. Elsevier, Dordrecht, 1992.

[90] K. L. Mittal, ed. *Polymer Surface Modification: Relevance to Adhesion, Part 1. Plasma Surface Modification Techniques.* VSP-International Science Publishers, 1995.

[91] D. T. Clark, A. Wilks, and D. Shuttleworth. "The Applications of Plasmas to the Synthesis and Surface Modification of Polymers," Chapter 9 in *Polymer Surfaces*, D. T. Clark and W. J. Feast, eds. Wiley, New York, 1978.

[92] R. E. P. Harvey. *Propagator Methods Applied to Long-Mean-Free-Path Plasma Chemistry.* PhD thesis, University of Wisconsin at Madison, 1995.

[93] R. E. P. Harvey, W. N. G. Hitchon, and G. J. Parker. *Journal of Applied Physics*, 75:1940, 1994.

[94] R. E. P. Harvey, W. N. G. Hitchon, and G. J. Parker. *IEEE Transactions on Plasma Science*, 23:436, 1995.

[95] V. P. McKoy, C. Winstead, and C.-H. Lee. *Journal of Vacuum Science and Technology*, A 16:324, 1998.

[96] J. W. Coburn and H. F. Winters. *Journal of Applied Physics*, 50:3189, 1979.

[97] J. W. Coburn and H. F. Winters. *Journal of Vacuum Science and Technology*, 18:349, 1981.

[98] U. Gerlach-Meyer, J. W. Coburn, and E. Kay. *Surface Science*, 103:177, 1981.

[99] R. A. Barker, T. M. Mayer, and W. C. Pearson. *Journal of Vacuum Science and Technology*, B1:37, 1983.

[100] S. C. McNevin and G. E. Becker. *Journal of Vacuum Science and Technology*, B 3:485, 1985.

[101] R. A. Rossen and H. H. Sawin. *Journal of Vacuum Science and Technology*, A 5:1595, 1987.

[102] D. C. Gray, I. Tepermeister, and H. H. Sawin. *Journal of Vacuum Science and Technology*, B 11:1243, 1993.

[103] M. E. Barone and D. B. Graves. *Journal of Applied Physics*, 78:6604, 1995.

[104] M. Balooch, M. Moalem, W. Wang, and A. V. Hamza. *Journal of Vacuum Science and Technology*, A 14:229, 1996.

[105] J. A. Levinson, E. S. G. Shaqfeh, M. Balooch, and A. V. Hamza. *Journal of Vacuum Science and Technology*, A 15:1902, 1997.

[106] J. P. Chang and H. H. Sawin. *Journal of Vacuum Science and Technology*, A 15:610, 1997.

[107] J. P. Chang, A. P. Mahorowala, and H. H. Sawin. *Journal of Vacuum Science and Technology*, A 16:217, 1998.

[108] H. Doshita, K. Ohtani, and A. Namiki. *Journal of Vacuum Science and Technology*, A 16:265, 1998.

[109] G. P. Kota, J. W. Coburn, and D. B. Graves. *Journal of Vacuum Science and Technology*, A 16:270, 1998.

[110] G. M. W. Kroesen, H.-J. Lee, and H. Moriguchi. *Journal of Vacuum Science and Technology*, A 16:225, 1998.

[111] S. Tachi, M. Izawa, K. Tsujimoto, T. Kuve, N. Kofuji, K. Suzuki, R. Hamasaki, and M. Kojima. *Journal of Vacuum Science and Technology*, A 16:250, 1998.

[112] K. H. R. Kirmse, A. E. Wendt, G. S. Oehrlein, and Y. Zhang. *Journal of Vacuum Science and Technology*, A 12:1287, 1994.

[113] M. J. Kushner. *Journal of Applied Physics*, 63:2532, 1988.

[114] M. J. McCaughey and M. J. Kushner. *Journal of Applied Physics*, 65:186, 1989.

[115] J. Perrin, M. Shiratani, P. Kae-Nune, H. Videlot, J. Jolly, and J. Guillon. *Journal of Vacuum Science and Technology*, A 16:278, 1998.

[116] H. F. Winters, J. W. Coburn, and E. Kay. *Journal of Applied Physics*, 48:4973, 1977.

[117] M. J. Kushner. *Journal of Applied Physics*, 53:2923, 1982.

[118] D. Edelson and D. L. Flamm. *Journal of Applied Physics*, 56:1522, 1984.

[119] K. R. Ryan and I. C. Plumb. *Plasma Processing and Plasma Chemistry*, 6:231, 1986.

[120] I. C. Plumb and K. R. Ryan. *Plasma Processing and Plasma Chemistry*, 6:205, 1986.

[121] E. Zawaideh and N. S. Kim. *Journal of Applied Physics*, 62:2499, 1987.

[122] E. Zawaideh and N. S. Kim. *Journal of Applied Physics*, 64:4199, 1987.

[123] P. Schoenborn, R. Patrick, and H. P. Baltes. *Journal of the Electrochemical Society*, 136:199, 1989.

[124] S. P. Venkatesan, I. Trachtenberg, and T. F. Edgar. *Journal of the Electrochemical Society*, 137:2280, 1990.

[125] M. Dalvie and K. F. Jensen. *Journal of the Electrochemical Society*, 137:1062, 1990.

[126] C. Szmytkowski, A. M. Krzystofowicz, P. Janicki, and L. Rosenthal. *Chem. Phys. Lett.*, 199:191, 1992.

[127] L. Boesten. *J. Phys. B*, 25:1607, 1992.

[128] T. Sakae, S. Sumiyoshi, E. Murakami, Y. Matsumoto, D. Ishibani, and A. Katase. *J. Phys. B*, 22:1385, 1989.

[129] A. Mann and F. Linder. *J. Phys. B*, 25:533, 1992.

[130] A. Mann and F. Linder. *J. Phys. B*, 25:545, 1992.

[131] H. F. Winters and M. Inokuti. *Physical Review A*, 25:1420, 1982.

[132] M. R. Bruce, Ce Ma, and R. A. Bonham. *Chem. Phys. Lett.*, 190:285, 1992.

[133] M. R. Bruce, Ce Ma, and R. A. Bonham. *Int. J. Mass Spectrom. Ion Processes*, 123:97, 1993.

[134] C. Ma, M. R. Bruce, and R. A. Bonham. *Physical Review A*, 44:2921, 1991.

[135] H. U. Poll, C. Winkler, D. Margreiter, V. Grill, and T. D. Mark. *Int. J. Mass. Spectrom. Ion Processes*, 112:1, 1992.

[136] T. Nakano and H. Sugai. *Japanese J. Appl. Phys.*, 31:2919, 1992.

[137] S. M. Spyrou, I. Sauers, and L. G. Christophorou. *J. Chem. Phys.*, 78:7200, 1983.

[138] I. Iga, M. V. V. S. Rao, S. K. Srivastava, and J. C. Nogueira. *Z. Phys. D*, 24:111, 1992.

[139] T. R. Hayes, R. C. Wetzel, and R. S. Freund. *Physical Review A*, 35:578, 1987.

[140] W. L. Morgan and A. Szoke. *Physical Review A*, 23:1256, 1981.

[141] F. A. Stevie and M. J. Vasile. *J. Chem. Phys.*, 78:5106, 1981.

[142] P. J. Chantry. *Applied Atomic Collision Physics, vol. 3, Gas Lasers*, E. W. McDaniel, W. N. Nigham, eds., Chapter 2, p. 35. Academic, New York, 1982.

[143] B. Peart, R. Forrest, and K. T. Dolder. *J. Phys. B*, 12:L115, 1979.

[144] M. Gryzinski. *Physical Review*, 138:A336, 1965.

[145] M. S. Huq, L. D. Doverspike, R. L. Champion, and V. A. Esaulov. *J. Phys. B*, 15:951, 1982.

[146] E. R. Fisher, M. E. Weber, and P. B. Armentrout. *J. Chem. Phys.*, 92:2296, 1990.

[147] B. Shizgal and A. S. Clark. *Chem. Phys.*, 166:317, 1992.

[148] L. D. B. Kiss and H. H. Sawin. *Plasma Processing and Plasma Chemistry*, 12:523, 1992.

[149] K. Miyata. *Studies on Fluorocarbon Plasma Chemistry in an Electron Cyclotron Resonance Etching System*. PhD thesis, Nagoya University, 1998.

[150] J. B. Adams and W. N. G. Hitchon. *Journal of Computational Physics*, 76:159, 1988.

[151] G. J. Parker, W. N. G. Hitchon, and D. J. Koch. *Physical Review E*, 51:3694, 1995.

[152] W. Y. Tan. *Journal of Applied Physics*, 79:3423, 1996.

[153] C. E. Wickershan. *Journal of Vacuum Science and Technology*, A5:1755, 1987.

[154] M. J. Cooke and G. Harris. *Journal of Vacuum Science and Technology*, A7:3217, 1989.

[155] A. Yuuki, Y. Matsui, and K. Tachibana. *Japanese. J. Appl. Phys.*, 28:212, 1989.

[156] R. N. Tait, S. K. Drew, T. Smy, and M. J. Brett. *Journal of Vacuum Science and Technology*, A10:912, 1992.

[157] D. Coronell and K. Jensen. *Journal of the Electrochemical Society*, 139:2264, 1992.

[158] T. Kinoshita, M. Hane, and J. P. McVittie. *Journal of Vacuum Science and Technology*, B 14:560, 1996.

[159] T. S. Cale, G. B. Raupp, and T. H. Gandy. *Journal of Applied Physics*, 68:3645, 1990.

[160] T. S. Cale and G. B. Raupp. *Journal of Vacuum Science and Technology*, B 8:649, 1990.

[161] T. S. Cale, T. H. Gandy, and G. B. Raupp. *Journal of Vacuum Science and Technology*, A 9:524, 1991.

[162] E. S. G. Shaqfeh and C. W. Jurgensen. *Journal of Applied Physics*, 66:4664, 1989.

[163] V. K. Singh, E. S. G. Shaqfeh, and J. P. McVittie. *Journal of Vacuum Science and Technology*, B 10:1091, 1992.

[164] J. I. F. Ulacia and J. P. McVittie. *Journal of Applied Physics*, 65:1484, 1989.

[165] S. Hamaguchi and M. Dalvie. *Journal of Vacuum Science and Technology*, A 12:2745, 1994.

[166] A. Rebei and W. N. G. Hitchon. *Physics Letters A*, 196:295, 1994.

[167] A. Rebei and W. N. G. Hitchon. *Physics Letters A*, 224:127–132, 1996.

[168] A. Rebei. *Generalized Random Phase Approximation: An Effective Action Approach*. PhD thesis, University of Wisconsin at Madison, 1998.

[169] J. C. Arnold and H. H. Sawin. *Journal of Applied Physics*, 70:5314, 1991.

[170] U. Kortshagen. *Journal of Vacuum Science and Technology*, A 16:300, 1998.

[171] H. C. Lin, C. C. Chen, C. H. Chien, S. K. Hsien, M. F. Wang, T. S. Chao, T. Y. Huang, and C. Y. Chang. *IEEE Electron Device Letters*, 19:68, 1998.

[172] S. Fang and J. P. McVittie. *IEEE Transactions on Electron Devices*, 41:1034, 1994.

[173] S. Fang and J. P. McVittie. *IEEE Transactions on Electron Devices*, 41:1848, 1994.

[174] K. P. Cheung and C. P. Chang. *Journal of Applied Physics*, 75:4415, 1994.

[175] K. P. Cheung and C. S. Pei. *IEEE Electron Device Letters*, 16:220, 1995.

[176] G. S. Hwang and K. P. Giapis. *Applied Physics Letters*, 71:1945, 1997.

[177] G. S. Hwang and K. P. Giapis. *Applied Physics Letters*, 71:2928, 1997.

[178] G. S. Hwang and K. P. Giapis. *Journal of Vacuum Science and Technology*, B 15:1741, 1997.

[179] G. S. Hwang and K. P. Giapis. *Journal of Vacuum Science and Technology*, B 15:70, 1997.

[180] A. K. Quick. *Electron Beam Neutralization of Large Aspect Ratio Features During Plasma Etching*. PhD thesis, University of Wisconsin at Madison, 1998.

[181] C. Jacoboni, C. Canali, G. Ottaviani, and A. A. Quaranta. *Solid State Electronics*, 20:77, 1977.

[182] P. D. Maycock. *Solid State Electronics*, 10:161, 1967.

[183] V. Axelrad. *International Journal of Numerical Modelling: Electronic Networks, Devices and Fields*, 2:31, 1989.

[184] V. Axelrad. *IEEE Transactions on Computer-Aided Design*, 9(11):1225, November 1990.

[185] M. Surendra. *Plasma Sources, Science and Technology*, 4:56, 1995.

[186] P. K. Moore, C. Ozturan, and J. E. Flaherty. *Mathematics and Computers in Simulation*, 31:325, 1989.

[187] W. M. Coughran, M. R. Pinto, and R. K. Smith. *IEEE Transactions on Computer-Aided Design*, 10(10):1259, October 1991.

[188] J. F. Bürgler, W. M. Coughran, and W. Fichtner. *IEEE Transactions on Computer-Aided Design*, 10(10):1251, October 1991.

[189] M. S. Adler. "A Method for Achieving and Choosing Variable Density Grids in Finite Difference Formulations and the Importance of Degeneracy and Band Gap Narrowing in Device Modeling." In *Proceedings of the Nasecode 1 Conference*, pp. 3–30, 1979.

[190] W. Cheney and D. Kincaid. *Numerical Mathematics and Computing*. Brooks/Cole Publishing Company, Pacific Grove, CA, 2nd edition, 1985.

[191] S. D. Conte and C. de Boor. *Elementary Numerical Analysis: An Algorithmic Approach*. McGraw-Hill, New York, 3rd edition, 1980.

[192] G. Dahlquist and A. Bjorck. *Numerical Methods*. Prentice-Hall, Upper Saddle River, NJ, 1974.

[193] C. S. Desai and J. F. Abel. *Introduction to the Finite Element Method: A Numerical Approach for Engineering Analysis*. Van Nostrand Reinhold, New York, 1972.

[194] E. P. Doolan, J. J. H Miller, and W. H. A. Schilder. *Uniform Numerical Methods for Problems with Initial and Boundary Layers*. Boole Press, Dublin, 1980.

[195] L. D. Kovach. *Boundary-Value Problems*. Addison-Wesley, Reading, MA, 1984.

[196] J. D. Logan. *Applied Mathematics: A Contemporary Approach*. Wiley, New York, 1987.

[197] J. M. Ortega and W. C. Rheinboldt. *Iterative Solution of Nonlinear Equations in Several Variables*. Academic Press, New York, 1970.

[198] R. D. Richtmyer and K. W. Morton. *Difference Methods for Initial-Value Problems*. Interscience Publishers, New York, 1976.

[199] G. D. Smith. *Numerical Solution of Partial Differential Equations: Finite Difference Methods*. Clarendon Press, Oxford, 1978.

[200] G. Strang and G. J. Fix. *An Analysis of the Finite Element Method*. Prentice Hall, Upper Saddle River, NJ, 1973.

[201] J. C. Strikwerda. *Finite Difference Schemes and Partial Differential Equations*. Wadsworth and Brooks/Cole, Pacific Grove, CA, 1989.

[202] M. N. O. Sadiku. *Numerical Techniques in Electromagnetics*. CRC Press, Boca Raton, FL, 1992.

[203] A. Thom and C. J. Apelt. *Field Computations in Engineering and Physics*. Van Nostrand, London, 1961.

[204] J. W. Eastwood. *Computer Physics Communications*, 43:89, 1986.

[205] J. W. Eastwood. *Computer Physics Communications*, 44:73, 1987.

[206] Ph. Belenguer and J. P. Boeuf. *Physical Review A*, 41:4447, 1990.

[207] A. Fiala, L. C. Pitchford, and J. P. Boeuf. *Physical Review E*, 49:5607, 1994.

[208] G. L. Tan, X. L. Yuan, Q. M. Zhang, W. H. Ku, and A. J. Shey. *IEEE Transactions on Computer-Aided Design*, 8(5):468, May 1989.

[209] J. P. Verboncoeur, G. J. Parker, B. M. Penentrante, and W. L. Morgan. *Journal of Applied Physics*, 80:1299, 1996.

[210] A. D. Boardman, W. Fawcett, and S. Swain. *J. Phys. Chem.*, 31:1963, 1970.

[211] Tran Ngoc An, E. Marode, and P. C. Johnson. *J. Phys. D*, 10:2317, 1977.

[212] D. Vender and R. W. Boswell. *IEEE Transactions on Plasma Science*, 18:725, 1990.

[213] M. Surendra, D. B. Graves, and I. J. Morey. *Applied Physics Letters*, 56:1022, 1990.

[214] M. Surendra and D. B. Graves. *IEEE Transactions on Plasma Science*, 19:144, 1991.

[215] V. Vahedi, G. Dipeso, C. K. Birdsall, M. A. Lieberman, and T. D. Rognlien. *Plasma Sources, Science and Technology*, 2:261, 1993.

[216] M. Surendra, D. B. Graves, and G. M. Jellum. *Physical Review A*, 41:1112, 1990.

[217] M. Surendra, D. B. Graves, and L. S. Plano. *Journal of Applied Physics*, 71:5189, 1992.

[218] K. A. Ashtiani, J. L. Shohet, W. N. G. Hitchon, G. H. Kim, and N. Hershkowitz. *Journal of Applied Physics*, 78:2270, 1995.

[219] W. N. G. Hitchon, T. J. Sommerer, and J. E. Lawler. *IEEE Transactions on Plasma Science*, 19:113, 1991.

[220] P. L. G. Ventzek, T. J. Sommerer, R. J. Hoekstra, and M. J. Kushner. *Applied Physics Letters*, 63:605, 1993.

[221] T. J. Sommerer, M. S. Barnes, J. H. Keller, M. J. McCaughey, and M. J. Kushner. *Applied Physics Letters*, 59:638, 1991.

[222] M. J. Kushner, W. Z. Collison, M. J. Grapperhaus, J. P. Holland, and M. S. Barnes. *Journal of Applied Physics*, 80:1337, 1996.

[223] T. J. Sommerer and M. J. Kushner. *Journal of Applied Physics*, 71:1654, 1992.

[224] H. D. Rees. *Solid State Commun.*, A26:416, 1968.

[225] U. Kortshagen. *Plasma Sources, Science and Technology*, 4:172, 1995.

[226] J. Bretagne, G. Gousset, and T. Simko. *Journal of Physics D: Applied Physics*, 27:1866, 1994.

[227] M. E. Riley, K. E. Greenberg, and G. A. Hebner, et al. *Journal of Applied Physics*, 75:2789, 1994.

[228] G. G. Lister. *Vacuum*, 45:525, 1994.

[229] F. E. Young and C. H. Wu. *IEEE Transactions on Plasma Science*, 21:312, 1993.

[230] G. G. Lister. *Journal of Physics D: Applied Physics*, 25:1649, 1992.

[231] H. Sugawara, Y. Sakai, and H. Tagashira. *Journal of Physics D: Applied Physics*, 25:1483, 1992.

[232] V. P. Konovalov, J. Bretagne, and G. Gousset. *Journal of Physics D: Applied Physics*, 25:1073, 1992.

[233] J. Yang, P. L. G. Ventzek, H. Sugawara, and Y. Sakai. *Journal of Applied Physics*, 82:2093, 1997.

[234] W. D. D'Haeseleer. PhD thesis, "Self-Consistent Bounce – Averaged Transport & Computations for Stellarators." University of Wisconsin at Madison, 1988.

[235] H. L. Berk and K. V. Roberts. *Methods Comput. Phys.*, 9:88, 1970.

[236] W. N. G. Hitchon, D. J. Koch, and J. B. Adams. *Journal of Computational Physics*. 83:79, 1989.

[237] T. J. Sommerer, W. N. G. Hitchon, R. E. P. Harvey, and J. E. Lawler. *Physical Review A*, 43:4452, 1991.

[238] G. J. Parker, W. N. G. Hitchon, and J. E. Lawler. *Physics Letters A*, 174:308, 1993.

[239] G. J. Parker, W. N. G. Hitchon, and J. E. Lawler. *Journal of Computational Physics*, 106:147, 1993.

[240] T. J. Sommerer, W. N. G. Hitchon, and J. E. Lawler. *Physical Review Letters*, 63:2361, 1989.

[241] W. N. G. Hitchon, G. J. Parker, and J. E. Lawler. *IEEE Transactions on Plasma Science*, 21:228, 1993.

[242] G. J. Parker, W. N. G. Hitchon, and J. E. Lawler. *Physical Review E*, 50:3210, 1994.

[243] V. Kolobov, G. J. Parker, and W. N. G. Hitchon. *Physical Review E*, 53:1110, 1996.

[244] G. J. Parker and W. N. G. Hitchon. *Japanese J. Appl. Phys.*, 36:4799, 1997.

[245] K. J. McCaughlin, S. W. Butler, and T. F. Edgar. *Journal of the Electrochemical Society*, 138:789, 1991.

[246] P. E. Riley, V. D. Kulkarni, and S. H. Bishop. *Journal of Vacuum Science and Technology*, B 7:24, 1989.

[247] R. J. Hastie, J. B. Taylor, and F. A. Haas. *Ann. Phys.*, 41:302, 1967.

Index